The Grand Science of Physics

Guide to the Amazing Universe of the Science of Physics

TIM VOK

The opinions expressed in this manuscript are solely the opinions of the author and do not represent the opinions or thoughts of the publisher. The author has represented and warranted full ownership and/or legal right to publish all the materials in this book.

The Grand Science of Physics
Guide to the Amazing Universe of the Science of Physics
All Rights Reserved.
Copyright © 2013 Tim Vok
v1.0

Cover Photo © 2013 JupiterImages Corporation. All rights reserved - used with permission.

This book may not be reproduced, transmitted, or stored in whole or in part by any means, including graphic, electronic, or mechanical without the express written consent of the publisher except in the case of brief quotations embodied in critical articles and reviews.

Outskirts Press, Inc.
http://www.outskirtspress.com

ISBN: 978-1-4787-1821-5

Outskirts Press and the "OP" logo are trademarks belonging to Outskirts Press, Inc.

PRINTED IN THE UNITED STATES OF AMERICA

"Physics is considered the most fundamental of sciences. It is about understanding Nature for her most basic rules, to see into the heart of reality, and to know the soul of the universe."
-physics seminar

"Climbing a mountain begins starting at its base and enduring the journey up. The moment when the climber reaches the peak and sits triumphantly at the summit is the great reward of climbing a mountain. Its because the mountain is there and also so because a climber exists wanting to climb the mountain that a mountain is climbed after all. Mt. Everest is the highest mountain in the world and there is no shortage of people who want to and have climbed this mountain. Sir Edmund Hillary and Tenzing Norgay are credited with climbing Mt. Everest. Physics knows its own Mt. Everest in the grand unified theory. No one as yet has climbed it, but history awaits those or the one who will do so."
-Voigt

"Physics had its origins in the magic practices of old. The ancients believed Nature was made from magic and hence was controllable by magic. Even though magic was discarded with time, the modern scientist follows in spirit what the ancient mages sought. The modern scientist through his mastery of science wields power over natural forces just as the ancient mages sought to do with spells, potions, and ceremony."
-physics club lecture

"In the beginning, God created the heavens and the earth…"
-Bible (Book of Genesis)

"If I have seen further it is by standing on the shoulders of giants."
-Isaac Newton

"Extraordinary claims require extraordinary proof."
-Carl Sagan

Dedication

This book had its inspiration and creation from college lectures in physics. It is a result of studying under physics professors and learning this amazing science. Thus this book is dedicated to the following people:

To physicists of any age.

To Albert Einstein.

"Science is for those who want to know better than superstition, to see further into Nature than others, and who want to understand themselves better than nothing at all."
-science lecture

Contents

Inspiration and Acknowledgements i
Preface: The Science of Physics iii
Introduction: Einstein's Dream (The Passion of Einstein) v
Outline vii
Chapter 1: Science 1
Chapter 2: Physics 53
Chapter 3: Physics Concepts 111
Chapter 4: Physical Fields and Forces 179
Chapter 5: Disciplines of Physics 221
Chapter 6: The Nobel Prize 325
Conclusion 354

"Modern physics is in some sense like sorcery. The ancients dreamed of wielding magic over natural forces. Although magic does not exist, the understanding of science does. It gives its student power over natural forces just as the ancients dreamed. It is in a small way the ability of Man at acquiring the power of God."

-physics lecture

Inspiration and Acknowledgements

This book began as student research into physics while in college. It is the result of long hours of reading, research, studying under physics professors, and working with physics professionals. It is the culmination of a long journey at trying to understand today's age of physics.

Today's physics world is alive with all kinds of new developments and discoveries. It is a complex fabric of theories and knowledge gathered in an attempt to understand the basic laws of the universe. It is an ever changing science filled with new chapters being written as this book goes to print. The future world of physics will seem scarcely like today's physics age centuries from now. Physics overall is a science that studies the nature of the universe. In ancient times, it would begin as a philosophical discussion into the nature of substances, rocks, time, and related qualities. Ages later it would advance into an epic science discussing space, time, mass, and energy. It would have many offshoot subjects and embrace ancient subjects like astronomy and cosmology. In the modern age, physics would grow into a labyrinth of a science difficult to comprehend and too vast for any physics student. This book in some small way tries to embrace the immensity of this science.

As with any labor of this magnitude, mistakes will inevitably result. Please forgive any omissions, errors, and distortions you may encounter. This work is the result of years of study and faults unfortunately are sure to happen. Any similarity to already published documents is entirely coincidental and does not represent an infringement.

Enjoy the experience of this book as it is a labor of love and hopefully gives you some insight into today's exciting world of science. Enjoy reading this book. Thank you and have a good life.

Tim Vok
October 2012

> "For some physics became a religion all its own, more a passion something of an obsession."
>
> -physics seminar

> "A journey of a thousand miles begins right where you stand."
> -ancient Chinese saying

Preface: The Science of Physics

Physics is a fascinating and magical science. It is in essence the study of how Nature works, behaves, and interacts. It is a quest to find its ultimate laws and to understand its bewildering diversity. Its study takes years of schooling and is only for the most ardent of students. It requires an extensive background in mathematics, chemistry, and other sciences to fully investigate. Few are the people who achieve a doctorate in its subject. The subject itself is filled with a complexity that defies any one thinker. Today, physics is too vast of science for anyone person. It is always changing with new ideas appearing all the time.

Today, physics is in an age of intense change and momentous events. Physicists worldwide are exploring all kinds of frontiers in the science. Issues range from superconductivity to quantum gravity with new developments appearing all the time. No one knows what tomorrow holds with this subject, but it is sure to be momentous as it was for thinkers before who conceived the quantum and relativity.

Physics had its origin in ancient times when primitive men began asking questions about how the world works. He did not have any answers, but took to contriving fables and fantasies and passing them off as "explanation" to his fellow tribesmen. From then on, various thinkers have come and gone who have advanced Man out of superstition into the "light" of today's age of physics. As of now, today's physics represents a collective achievement of so many thinkers who have contributed something here and there to help men understand the world in which he lives.

Physics represents a kind of modern sorcery. As ancient "wizards" sought the use of magic to control elemental forces, today's physicists wield similar power, but call their "magic" electricity, gravity, nuclear power, and natural forces. Although physics has long shed its connotations with magic, it is today a thriving and professional discipline of science exploring how the universe works.

As of now, physics is in the midst of its gravest age of crisis. It is trying to understand the basic laws of the universe, yet it is in a state of chaos, bewilderment, and impass as it cannot find the true, basic, and fundamental understanding of creation. How this crisis resolves itself and the quest to solve this remains for a future time. In this book, we will review discussions and lectures on the amazing science of physics.

"The incomprehensible thing about the universe is that it is comprehensible."
-Albert Einstein

"I want to know God's thoughts. The rest are details."
-Albert Einstein

Introduction: Einstein's Dream (The Passion of Einstein)

Albert Einstein, the legendary thinker and titan of 20th century physics had a curious and profound dream he bequeathed to physics. He dreamt that the universe obeyed a single master theory of physics with its own set of fundamental equations. This theory would have a single master equation, a basic principle, the ability to derive mathematically any other physics theory, and an "explanation" for Nature's laws and diversity. He sought to find this theory and spent years fruitlessly searching. However, he had a deep and abiding confidence bordering on physics faith that this theory did exist. However, he died in 1955 failing to achieve this one last service to physics.

This mythical physics theory Einstein so feverishly believed existed is today called by many names. Examples are the master theory, the super theory, the grand unified theory, the one theory, the master theory, the fundamental theory, the unified theory, the unified field theory, the God theory, the theory of everything, the theory of the universe, and last but not least the theory of the universe. Its existence is a kind of fantasy in physics, a dreamy thing many believe exists, but no one can "prove" exists.

Today, physicists the world over are searching for this theory, the grand unified theory. It is the passion of thousands of thinkers who labor to solve the mystery that even the great Einstein could not solve. It represents in some sense physics' equivalent to the Holy Grail, a magical treasure waiting for some hero to stumble upon. As of now, its issue is discussed in all kinds of places ranging from colleges to corporations. As of yet no one has found it and it is quite possible it does not exist. However, this does not stop the worldwide 'army' of researchers, hobbyists, and physicists from trying to find it.

For the time being, the quest to find the grand unified theory represents a kind of scientific treasure hunt. Its prize is none other than the discovery of the basic laws of the universe and insight into why Nature is what she is. The journey to find this theory goes on and nobody knows how it will end nor with whom. It represents an achievement that may forever remain unsolved and if that is so, represents a kind of futility that scientists for ages will dwell upon. However this adventure finishes, it represents in some sense a last, great, and ultimate fantasy journey to end the journey begun by our primitive ancestors ages ago. For now, the grand unified theory remains just a fantasy, somewhere in the netherworld of the unknown of science waiting for its champion to find this prize.

To whosoever solves this greatest of physics mysteries, a special place in the science of physics will go. Such a person will join the ranks of Newton and Einstein as some of physics' greatest thinkers. Nobody knows who will live this future chapter of physics, but it is sure to be one of the great stories of history and perhaps the greatest story of physics history.

"What does religion mean? What is its truth really?"
-class discussion

"At the present rate of progress, it is almost impossible to imagine any technical feat that cannot be achieved – if it can be achieved at all – within the next hundred years."
-Arthur C. Clarke

Outline

This book is meant to be a discussion on science and physics. This book is divided into the following sections with their discussions:
- Chapter 1: The world of science. This chapter is given for the reader to have a basic familiarity with science. It will discuss the scientific method and related issues.
- Chapter 2: The world of physics. This chapter is given for the reader to have a basic familiarity with physics.
- Chapter 3: Discussion of concepts. Physics requires an education on some very basic concepts to more ably understand the subject. Physics discusses seven basic ideas in mass, energy, mass-energy, space, time, spacetime, and mind to more accurately understand.
- Chapter 4: Discussion on forces. Physics presently knows four fundamental forces and other issues, brief discussions on them are given for the reader. They are called by the names of gravity, electromagnetism, the strong nuclear force, and the weak nuclear force.
- Chapter 5: Discussion on the many branches and sub-disciplines of physics.
- Chapter 6: Discussion on the Nobel Prize for Physics.
- Concluding statements.

"For so long superstition guided men in how they understood Nature. However Nature would show that superstition was inadequate and in reality wrong. A new kind of thinking would replace superstition and this was science. Science offered a light in contrast to darkness."

-physics seminar

To begin with in our discussion and exploration of the science of physics, we shall first begin with a discussion on the parent subject of physics in science. It is with science that physics gains its teaching, worldview, nature of thinking, and direction. Let us now review a discussion on science.

Chapter 1
Science

"Science is about understanding Nature better than superstition. People of science see the world through effects, natural laws, equations, order, and phenomena. People of superstition see the world as actions of spirits, magic, the gods, demons, angels, and myths. Human history has known both kinds of people, but it is known people of science understand the world better than people of superstition."
                                                            -class discussion

Science

"Science is for the person who wants to know the secrets of the universe."
                                                            -class discussion

Science is a grand drama of Man. It began ages ago with primitive men asking simple questions about Nature. In time, it became intertwined with superstition and has stayed that way for ages. Since then, it has changed and evolved studying and dissecting Nature's myriad features. It has grown into various branches called physics, astronomy, chemistry, biology, geology, and so on. It would have its formal beginnings in the time of the 14th to 17th centuries in an era called the Scientific Revolution. In this time, thinkers believed superstition and a subject called natural philosophy would define the intellectual world of this time. However dramas would abound that suggested superstition would not work as a way of explaining Nature. In this ferment, a new kind of thinking would be born and this would be called science. Today science would evolve and advance so fantastically as to shape the entire world. It now pervades the world in such things as space travel, electronics, technology, and much more. It is indispensable to the modern age, shaping, changing, and making it what it is. We now begin with a discourse on this subject so vital to understanding the modern age.

"The whole purpose of science is to understand Nature and her many features."

-Voigt

The World of Science

Science is a curious word. It refers to a vast body of subjects that seek to study the various aspects of Nature. Aspects of Nature are the following: heat, light, life, intelligence, society, Man, animals, plants, stars, atoms, and so much more. For each subject there is an appropriate science. Examples are botany for plants and optics for light. Science is about learning knowledge, knowledge about a subject. The scientist wants to know the rules of the subjects, analyzes its behavior, and creates theories to describe its behavior. As more knowledge is discovered about a subject, the knowledge of science grows. Science is taught worldwide and is essential to modern civilization. Science has its system of thinking about how to explore Nature and this is known as the 'scientific method'. Today science is a vast labyrinth of issues, findings, topics, and discussions found almost everywhere. It is now so vast it seems no one can keep up on it anymore.

What is Science?

Science is a system for investigating something about Nature, a means to learn about a natural quality by the use of reason, thinking, investigation, experiment, and other tools. It attempts to understand a natural quality with intelligence and tools.

Before Science

Before modern science evolved into existence, science-like thinking was practiced for centuries. As to who was the first to ask questions about the universe or formulate the first "explanation" for Nature, nobody will ever know. However, he or she began the journey of Man at trying to invent that later subject known as science. In ancient times, people believed in all sorts of things. Examples are the gods, magic, potions, sorcery, witchcraft, astrology, alchemy, and a whole host of other issues. Beliefs in such things passed for "understanding" of how Nature worked and in some sense represented Man's views of "science" for countless ages. Primitive men believed that natural forces were controlled by capricious, powerful entities called the gods. The gods were described as "superhuman" possessing great knowledge and power and being somehow above Man. Men took to worshiping the gods and dedicated many temples and religions to them. Whole civilizations were founded to honor the gods and draw inspiration from "their" guidance. In time, various thinkers would come who would question the beliefs in the gods. In time, this began the historical trend of disbelieving in the gods and the erosion of superstitious beliefs about Nature. Later on, thinkers like Newton, Copernicus, and Galileo would work to found the Scientific Revolution and cause Man's break with superstition in trying to understand how Nature worked. Science in some way had its beginning in people who thought superstition was nonsense and was not capable of explaining the natural world.

Beginnings of Science ("Scientia", Latin to "know")

Science had its origins in such things as idle curiosity on natural things, experimenting with natural features like fire and rocks, and inventing stories to "explain" natural things. This has spawned movements among primitive men that can be called superstition. Such things like beliefs in magic, ceremonies, rituals to appease the gods, incantations, and the making of potions all illustrate this activity. These practices continued for ages being believed in by countless societies. Even now, such practices survive in more primitive parts of the world. However its from these practices that science would evolve and change the world later on.

Superstition

The ancients believed that the world was made from magic, that spirits and gods existed, and that the world was filled with all manner of supernatural creature and thing. Out of an environment like this, beliefs would arise that would become the first "natural beliefs". The ancients thought spells could cure the sick, that items like amulets could ward away "evil spirits", and that a curse could be placed on people. In all, the ancients believed in many kinds of superstitious beliefs like the following:

- curses –hexes –spells –incantations –angels –demons –supernatural animals –werewolves –vampires –the gods –magic potions –the Holy Grail –elixirs of immortality –faith healing –healing waters –goblins –ghosts –hauntings –seances –omens -ghouls -black magic -witches -prophecy -Heaven -Hell -Satan

Science Arises From Superstition

Many sciences are known to have originated in ancient superstitious practices. Many examples known to history are the following:

- Alchemy gives birth to chemistry. Alchemy was an attempt to change lead into gold for which the ancients prized. It however would lead the ancients to probe natural substances like soil, rocks, gems, minerals, woods, and other things. In time this would lead to the concept of 'element' and later chemistry.
- Astrology leads to astronomy. Astrology is a superstition where the movements of stars and planets could somehow influence people. It would inspire ancient peoples to observe the sky and thus lead to the science of astronomy. In time astronomy would emerge from astrology.
- Physics arises from natural philosophy, Aristotelian thinking, and Greek ideas.
- Medicine arises from shamanistic practices of herbalism and medicinal treatment.
- Geometry arises from Pythagoreanism and the mystical reverence for shapes.

SCIENCE

- Biology arises from shamanistic practices into life.
- Numerology, primitive counting, and reverence for shapes would lead to mathematics.
- Ancient Earth cults would give rise to geology. The ancients would see such things as volcanoes, quakes, geysers, and other terrestrial dramas as actions of the gods. This would inspire them to investigate them and thus found geology.
- Ancient superstitions into gem minerals would lead to gemology.
- Shamanistic practices to "cause" rain to fall would give rise to meteorology.

In all many sciences would emerge from some superstitious practice from ancient times. However not all sciences would do so, others would emerge from the modern world or be a blend of sciences themselves. Please read many books on mythology, fable, and fantasy.

Simple Questions

Man has always had a fascination with the natural world. He has taken to constructing statements of words beginning with the basic question words: Who, What, Why, How, When, and Where. Statements beginning with these words are called questions. Questions form the practice of inquiry or looking into things. Seeking answers to questions is an activity that would lead to science.

The Nature of Science

Science is an intelligent activity that engages in these dramas:
- Studies an aspect of Nature methodically with or without the use of mathematics.
- Searches for knowledge, order, and understanding about a natural quality.
- It asks questions, composes hypotheses (claims), and conducts experiments to test claims.
- It will create categories to classify things.

- It allows for discourse, criticism, challenge, and investigation into claims.
- Has a community of thinkers who study the science and govern its organization.
- Publishes findings in a journal for other science thinkers to know about.
- It can compose laws and theories about how a natural quality behaves.
- It can grow in its knowledge being advanced by the contribution of laws, methods, technology, texts, findings, and other drama.
- Organizations may be founded to govern how a science is taught, what a science's standards are, and other features.

Please read about the nature of science.

Things and Phenomena

Nature is filled with countless things or objects. Examples are planets, stars, trees, animals, plants, water, rocks, mud, volcanoes, gems, birds, cats, shoes, houses, people, dirt, wood, substances, chemicals, and so much. Each kind of thing has been investigated, analyzed, and conceived as to being the "subject matter" of a science. If there were no things there would be no science. Many a science is named by the convention of having a word denoting some thing and then adding suffixes like -ology (study of), -ometry (the measure of), -ics (discussion of), -istry (study of), -ography (shape of), -omics (nature of), -osophy (knowledge of), or -onomy (information about). Thus many sciences gain their name like the following:

Psyche (mind) + -ology = Psychology
Geos (Earth) + -ography = Geography
Audio (sound) + -ology = Audiology
Geos (Earth) + -metry = Geometry
Bacteria + -ology = Bacteriology

SCIENCE

Logos and Gnosis

These are ancient words known to the Greeks, Romans, and societies after them. They literally mean order and knowledge. It was the task of early scientific thinkers to try to devise a system of order and collect knowledge on natural features. They had studied plants, animals, rocks, bones, and other features of Nature. They tried to found communities to discourse and "explain" natural features. In time, their research would lead to the founding of sciences like biology, physics, chemistry, geology, and so on. Its from here that science begins to "grow up" and appear in the world as the modern movements they are. Logos would in time become the word fragment –ology added to words like bio- (life), geo- (Earth), and other root words to invent the science words of biology, geology, and so on. Gnosis would find itself incorporated into the word, Gnosticism denoting educated, learning movements of ancient times.

Please read about Greek philosophy and the nature of these and other ancient words.

Observation

Observation is the act of sensing something and saying it exists. Examples are seeing clouds in the sky, smelling flowers for a scent, tasting food, touching things, and hearing sounds. Scientists try to observe all manner of natural feature by their senses to determine if something exists. Whenever a new thing is discovered in Nature, it is said to be observed by someone. When many scientists all claim something has been observed, it is said to exist. Observation underlies every science as it is the activity of a scientist to determine if something exists or not.

Hypothesis and Experiment

In science there are two words that occupy the attention of researchers more than any other. They are called hypothesis and experiment. A hypothesis is a claim about a natural thing. It can be like saying "the sky is blue" or "water is wet". It is a claim that says a natural thing has a property, a quality, or a "nature" of some sort. An experiment is a test of a hypothesis. It is an "act" to see if some claim is "right" or not. An experiment can be to see if light goes at only one speed or if gravity pulls down. Both these words are vital to the activity of science and of science overall.

Exploration and Discovery

To ancient peoples, the world was an unknown and bewildering place. It was filled with distant lands, strange animals and plants, unknown societies and tribes, and many other features. Various people arose through the ages who would challenge the unknown and explore the world. They collected plant specimens, visited volcanoes, gazed at the stars, analyzed animal remains, and dug in the ground. Its from these ancient activities that the first scientist explorers would arise. Their activities continue in spirit in the form of modern science research.

Please read about the many adventures of science in ages past. Examples are Darwin's journey, the exploration of the North and South Poles, the voyages of Columbus, Captain Cook's voyages, Lewis and Clark, Cortes, Pizarro, and so on.

Investigation by Science

Scientists seek to analyze and understand the many features of the natural world. They may partake in the following activities:

- Collect samples. Scientists will attempt to collect "samples" or specimens of something being studied. Samples can be leaves, carcasses, cells, rocks, minerals, wood specimens, blood samples, gas samples, evidence of particles, and so on.
- Describe samples. Scientists will comment on the qualities of a sample. They may weigh, measure, identify features, catalog, observe, and so on.
- Construct a picture. Scientists will try to "comprehend" the whole reality of a sample.
- Construct a theory. Scientists will construct an "explanatory picture" or theory describing the sample and how it fits in a greater "whole".

- Use tools. Scientists employ many tools to engage in the activity of science. Tools can be scales, flasks, bags, measuring devices, calculators, catalogs, and so on.

Science is an elaborate and complex activity performed by trained scientists.

Divisions of Science

Science has been sub-divided into various "fields" of science. Presently these are the known kinds of science fields:

- Health sciences are subjects like medicine, cardiology, healing, nursing, etc.
- Natural sciences are subjects like physics, chemistry, and geology.
- Cognitive sciences are subjects like psychology, psychiatry, and sciences of mind.
- Agricultural sciences are subjects like dairy science, ranching, and so on.
- Social sciences are subjects like political science, criminal justice, and anthropology.
- Formal sciences are subjects like philosophy, mathematics, logic, and so on.
- Military sciences are subjects like police science, naval engineering, and so on.

Kinds of Science

Various thinkers would come and "categorize" natural features. This has lead them to found different kinds of science to study various features of Nature. Examples of sciences now taught worldwide are the following:

- Zoology: the science of animals.
- Cryptozoology is the study and search for animals rumored to exist in legend, myth, and account.
- Anatomy is the study of body parts and features of plants and animals.

- Ornithology is the study of birds.
- Parasitology is the study of parasites and is also a branch of medicine.
- Virology is the study of viruses (virii).
- Bacteriology is the study of bacteria.
- Anthropology is the study of Man, this has also been called 'hominology' as well.
- Primatology is the study of primates or apes and monkeys.
- Botany: the science of plants. Herbalism has also been used to name botany as well.
- Entomology: the study of insects.
- Apiology is the study of bees and bee-keeping.
- Cetology: the study of whales.
- Mammalogy: the study of mammals.
- Herpetology: the study of reptiles.
- Indology is the study of India.
- Mythology is the study of ancient stories, fables, myths, creation stories, and related.
- Mathematics: the science of logic, order, number, shape, and harmony.
- Physics: the science of natural forces.
- Cosmology: the science of the universe.
- Selenology is the study of the Moon.
- Astronomy: the science of the stars and celestial objects.
- Biology: the science of life.
- Library science is the science of how to run, order, and maintain libraries of books.
- Chemistry: the science of chemicals.
- Thermodynamics is the study of heat.
- Numismatics is the study of coins, tokens, paper money, bullion, and currency.
- Psychology: the science of intelligence, brain, knowing, and behavior.
- Economics: the study of money, finances, and related issues.

SCIENCE

- Seismology is the study of earthquakes, waves, and the like.
- Arachnology: the study of spiders and scorpions.
- Theology: the study of God and religious movements.
- Sinology: the study of China.
- Nippology is the study of Japan.
- Sociology: the study of societies of Man.
- Geometry is the study of shapes and forms and is also a branch of mathematics.
- Geography: the study of maps.
- Geology: the study of the Earth.
- Planetology (planetary science): the study of the planets.
- Computer Science: the study of computers.
- Political Science: the study of societies, politics, and related issues.
- Egyptology: the study of Egypt (more specifically Ancient Egypt).
- Hittitology: the study of the ancient Hittite empire.
- Assyriology: the study of the ancient Assyrian empire.
- Optics: the study of light.
- Cardiology: the study of the heart and also a branch of medicine.
- Meteorology: the study of the weather.
- Aeronautics is the study of airplanes, flying, and related issues.
- Space Science is the study of space travel, astronauts, space physics, astronomy, etc.
- Military Science: the study of war, armies, weapons, strategy, tactics, and so on.
- Police Science: the study of crime, police procedures, arrest situations, and so on.
- Fire Science: the study of wild fires, arson, firefighter methods, and so on.
- Medicine is the study of curing people and animals from diseases, injuries, infirmities, and other health conditions.
- Native American Studies is the study of the many kinds of Indian tribe.

- History: the study of past events, dramas, characters, themes, and so on.
- Speleology: the study of caves, cave systems, and cave exploration.
- Polar research is the study of polar regions.
- Ufology is the study of the UFO phenomena.
- Oceanography is the study of oceans, seas, and their many issues.
- Vexillology is the study of flags and banners.
- Ecology: the study of life, ecosystems, the environment, sustainable balance, etc.
- Glaciology is the study of glaciers, ice sheets, and so on.
- Toxicology is the study of poison.
- Archaeology is the study of history, artifacts, ancient societies, and related issues.
- Arcology is a blend of architecture with ecology.
- Architecture is the design of constructions.
- Engineering is the activity of making constructions from designs of architecture.
- Taxonomy is the study of ways to classify life or construct a system of life relations.
- Volcanology is the study of volcanoes.
- Paleontology is the study of fossils, ancient life, and so on.
- Parapsychology: the study of the paranormal. This subject however is embroiled in controversy as it is not known to study anything "real". Its history has had it at times been recognized as a science and also as a pseudoscience. It is today, a curious subject struggling to gain acceptance as a science, but having to endure its share of ridicule for being questionable and uncertain of reality.
- Geophysics: a blended science of geology and physics.
- Biophysics: a blended science of biology and physics.
- Astrophysics is a blend of astronomy with physics.
- Genetics is the study of genes, DNA, cloning, the genome, and so on.

SCIENCE

There are in reality many kinds of science each with its practitioners, journals, communities, and colleges dedicated to its study. Sciences continue to be invented all the time. As people realize there are new frontiers to Nature to study or that two sciences can be combined into one, science grows and expands to become an evermore complex subject.

Science Cross Disciplines

Many times various sciences "share" a "common knowledge base". That is two separate sciences have qualities in common with each other. For example both mathematics and physics share the use of numbers. Physics and chemistry share in the study of chemicals (mass objects). Biology and physics share in the study of life (mass objects that are alive). Many times sciences are "combined" into hybrid sciences which study common knowledge between them. Examples of such hybrids are geophysics, biophysics, biochemistry, physical chemistry, mathematical physics, and so on. There are in reality many kinds of combined science and more can be devised.

Pure, Applied, and Experimental Science

Science is a vast subject where people learn about the knowledge of Nature and use it for their own purposes. Science has been subdivided into various activities of how it is used. Examples are:

- Pure science: this is science done by pure thinking and research.
- Applied science: this is the application of science for technology and human use.
- Experimental science is the "physical testing" of theories to determine reality.

"Truth is sought for its own sake. And those who are engaged upon the quest for anything for its own sake are not interested in other things. Finding the truth is difficult and the road is rough." -Ibn al Haytham (midieval Arab scholar)

Knowledge and Truth (What is right?)

Knowledge refers to such things as concepts, data, facts, findings, theories, and other bits of "data". Truth refers to what is right, correct, real, and the opposite of false and lie. These two words are essential to the modern scientist as he attempts to know both of these things. Science is so deeply intertwined with these words that the reader is encouraged to explore their intricacies in works of philosophy. Today's scientist is on the "hunt" for more knowledge to compose what is called a "theory". A theory is a body of beliefs that attempts to "explain and/or describe" a natural thing or behavior and predict future behavior in as simple of terms as possible (Ockham's razor). The scientist in his investigation of phenomena (or natural occurrences) endeavors to question, criticize, describe, explain, and doubt what he is "observing" or sensing. It is his pursuit of "truth" that draws him to investigate natural things. When he believes he has found something that time and again refuses to be disbelieved, debunked, or discarded, he comes to the realization that he has found a truth of Nature. A truth of Nature is later called a fact and takes its place within the "canon" of a science or its "body of beliefs". In this the scientist is always searching for truth and that next truth to add to a science.

Reproducibility

At times in the history of science, scientists will announce the discovery of something that will eventually be called facts. They will make claims that something is real, publish their findings in a respected journal, and go on to claim that a science is advanced. Many times, scientists err and announce the discovery of things that are not real. When this happens, fellow scientists will conduct experiments to test for these claims. If these other scientists end up claiming the discovery of something new like the previous scientists, it is said that a discovery is real. This overall is called reproducibility and is a method for scientists to "verify" that something has in fact been discovered.

SCIENCE

Credit ("Credit is given where credit is due." – popular saying)

Credit is the idea of commending, "glorifying", citing, or acknowledging scientists who have found something new in Nature. Scientists are given credit in journals for finding new facts or theories in a science. Sometimes the new finding is named in honor of the scientist or scientists who found something. Examples of credit are allowing discoverers to name new elements or astronomical objects, to name new effects after their discoverers, and to name new laws or constants (numbers) after their discoverers. If discoveries prove to be monumental, fantastic, impactual, or great, then the discovering scientists may gain awards like the Nobel Prize, be entered into history, or gain fame for their achievements.

Consensus

Consensus refers to various scientists agreeing on an observation, a result of an experiment, or a discovery. It refers to agreement among parties that what is said is in fact true. Examples of consensus are congressional votes, jury rulings, panel rulings, and so on. Consensus underlines science and its activities.

An Example of a Theory

Theories are creations fundamental to any science. They are bodies of beliefs that attempt to explain and describe how a natural phenomenon behaves. Examples of famous theories are the theory of evolution, quantum mechanics, general relativity, the BCS theory, and so on. A good example of the nature of a theory is the following: Consider the discovery of a broken window and a ball found inside a house. An investigator into this affair would find broken glass, a hole in the window, and other qualities called "data". The investigator would devise a "theory" or explanation (description that explains) that would go like this: somebody threw a ball and broke a window leaving a hole in the glass. The ball itself lies inside the house. This represents in some way a "theory" of this drama or an explanation that someone caused this drama to happen. Taken in

THE GRAND SCIENCE OF PHYSICS

stride, this represents what an explanation in science is and is thus a theory. Science is filled with many kinds of theory and they each have their power in explaining natural behaviors. Theories are important to science and it is the task of the scientist to compose them in order to advance a science. If a theory does not account for the "facts" of a drama or case, it is replaced or discarded in order to find a better theory. Facts not in agreement with established theory suggest a scientific theory is wrong and needs revision. Crises in science are where theories cannot account or explain the facts of a drama. Today, science is filled with numerous dramas where facts just cannot be explained and this represents a frontier in science. From here, science discovery can proceed and hopefully, new theories can be invented to explain natural phenomena once and for all.

To understand the nature of theories more closely, the reader is encouraged to study crime events and attempt to explain them in some way based on the nature of the evidence. Learning to solve murder cases is a good way to explore what a theory is. A court trial is a "legal drama" that composes two theories about an event thought to be a crime. One side called the prosecution says a crime took place and another side called the defense says a crime did not take place. Both sides present "evidence" as to what occurred at an event thought to be a crime. Both sides have their "theories" about how to explain the event thought to be a crime. How a jury (or panel of citizens) rules is considered to be an "answer" as to which side's theory is right or wrong. If a jury rules guilty, the prosecution's theory is said to be "right". If a jury rules innocent, the defense's theory is said to be "right". The claim that a crime has been committed is a hypothesis. Evidence presented in court represents "data" or knowledge. Each side's claim as to what happened at an event is a "theory". The jury's ruling is an experiment to test for truth to see which theory is "right".

Theory Faith

A theory is a description and explanation all rolled up into one or a unity. It is a vision of how to explain natural things. If a theory accounts for phenomena simply and beautifully, it is then "trusted" by scientists who believe it is "right" or the "truth". This situation remains for years and can be looked upon as "faith" or "deep trust" in science. However, if some fact should come along that says a theory is "false" or "wrong" or unable to explain phenomena then it is considered challenged of authority. In time as more facts are learned, a theory may lose all "faith" by scientists and be discarded. What happens is that this theory is replaced with another and the "old theory" is considered "overthrown". Overthrowing theories in science is a rare occurrence, but when it happens it can be considered momentous and revolutionary.

Ockham's Razor (Parsimony)

This is a belief credited to William of Ockham. It discusses a notion that the "best answer is the simplest answer" or that the "theory" that is most simple is the "right" theory of a phenomena. An idea of this is to see a window with a hole in it. A "simple theory" of this would be to say a bullet or a ball passed through the window. This theory is "simple" in that a hole implies an object (not necessarily a bullet, but it could be a rock or a baseball) passing through the window causing the hole. Ockham's razor is an interesting discussion and is used widely in science.

Please read further on Ockham's Razor.

Overthrowing Theories

From time to time, theories in science will attain to a deep trust among scientists. They will believe their views and beliefs in a theory so strongly that nothing will ever come along to challenge trust in a theory. However, sometimes new facts are discovered that suggest a theory is not right. When this happens, thinkers may come along and revise or challenge a theory in science. When this happens, its thought a momentous drama is occurring where a theory is in the process of being "overthrown" or shown to be questionable of authority. Dramas in history where "scientific theories" have been thrown into question

and hence overthrown (or replaced) are the following:
- The gods. Ancient peoples believed "superior intelligences" called the gods existed and governed Nature by magic and their understanding. Belief in the gods represents a kind of theory as to how Nature is to be explained. Later thinkers would question the reality of the gods (Thales of Miletus) and show the gods do not exist. This would have the effect of overthrowing a theory that says the gods existed.
- Ether. Thinkers for ages believed that a substance called "ether" (liquid reality) existed. It pervaded the universe, allowed for waves of light to pass, and existed almost supernaturally, but without any proof. Two thinkers named Michelson and Morley would perform a famous experiment in the 19th century looking for ether. They could not find it and this overthrew one of physics' most cherished beliefs. Later on, this finding would be used by Einstein to discover relativity.
- Four fundamental substances. In ancient Greece, various thinkers thought Nature consisted of just four "elements" in Earth, Air, Water, and Fire. They were thought to be fundamental, irreducible, not made of anything else, and unique. Various thinkers who came after (chemists) analyzed these elements and realized they could be broken down into other substances. Water for example was found to be made of the gases, hydrogen and oxygen. This and other findings disproved the various beliefs in the fundamental substances and hence these Greek ideas are thought overthrown.

Anomaly

Sometimes a discovery appears that challenges how scientists think about Nature, a theory, or a belief. This is referred to as an "anomaly" or curiosity that does not fit in the scheme of things. Scientists will investigate an anomaly and attempt to explore it for the "good" of science. Exploring anomalies can lead to discovery.

SCIENCE

The Scientific Method

There was a time when men knew almost nothing about Nature. He was ignorant, superstitious, and prone to fantasy. He invented all kinds of explanations for the bewildering world in which he lived. In time, various thinkers would arise that would question these explanations. Eventually, these thinkers would believe that perhaps magic did not exist and that the gods were not a reality. This began a process whereby people began to question Nature and sought explanations that did not involve superstition. This process would culminate in later ages in the creation of the Scientific Method, a formal body of beliefs by which natural phenomena (natural features) could be investigated. The Scientific Method has these qualities as aspects of its discussion:

- Phenomena. This is a word pertaining to a vast assortment of natural behaviors. Examples are heat, light, gravity, motion, water flow, rotation, burning, melting, freezing, thawing, atoms, fire, lightning, thunder, sound, aurora, quakes, twisting, and so on. The scientist will explore the nature of any of these things using the resources of science to help him understand what it is.
- Be curious. It is because Nature is filled with what is unknown that a scientist gains his mission in exploring his curiosity with things.
- Exploring. Nature to ancient men was a vast unknown. Many thinkers would explore its domains and from this identify issues to research.
- Identify a phenomenon. This refers to naming, identifying, or discussing just what is to be studied in a science.
- Questions. The activity of asking questions or sentences of inquiry into phenomena. Questions usually begin with the fundamental words of language: Who, What, Why, When, Where, and How. Examples of questions are "What is light?", "How does gravity operate?", and "Why do things exist?"
- The search for answers. The activity of trying to answer natural questions.

THE GRAND SCIENCE OF PHYSICS

- Conceptualize. The scientist will invent models, words, discussions, and scenarios by which to describe a phenomena.
- Seek after truth. Truth in science refers to what is right, real, proveable, and existent. It is the goal of science to discover the truth.
- Making claims (called hypotheses). The activity of stating something about a phenomenon. An example is light has component colors like red and blue. A scientist will make up a claim and then leave it to experiment to test that it is true or not, real or not.
- Devising experiments. The activity of testing claims or performing experiments. An experiment is a test of truth and existence. Experiments are complex dramas that can range from sending probes into space or dropping probes into the sea to "sense" for some data. Famous experiments of science lore are the Michelson-Morley experiment, gold foil experiment, Eddington-1919 drama, and so on.
- Data refers to facts, discoveries, readings, and other things counted as information. Scientists seek to gather data to analyze, compose theories, and establish that something exists or not.
- Gathering data. This is the activity of gathering knowledge about a phenomenon useful in theorizing later on. Expeditions may be organized to journey to places to gather data.
- Describing. This is the activity of observing a phenomena and then commenting on its qualities. An example is to say a cloud appears white, flies in the sky, emits rain, crackles lightning and thunder, and so on.
- Explaining. This is the activity of discussing how something works or how it originated.
- Probe and Dissect. This is the activity of investigating intensively and intrusively into a thing.
- Criticisms. The activity of "doubting" claims and requesting the presentation of proof or evidence that the claim is right.

SCIENCE

- Skepticism. At times scientists make boastful claims that sound like fantasy more than science. In this they get doubted by other scientists and thus doubters are 'skeptical'. If a claim however farfetched should prove real, it will eventual win out and silence skeptics.
- Research. The activity of probing or investigating a phenomenon. It can also refer to doing library research to learn more about an issue.
- Theorizing. The activity of trying to invent a theory to explain phenomena. An example is light originates from the "burning" of a star or ice is "frozen water". Theorizing is a main activity of science as the scientist attempts to explain a thing or phenomena.
- The founding of theories. The activity of inventing a theory that can withstand any criticism as well as explain and/or describe all facts. Theories that do a great job of explaining natural phenomena are eventually called science teaching. Theories remain the main body of teaching about a science until someone presents knowledge that perhaps a theory is not "right". When this happens, a theory is said to be overthrown and is in need of replacing. When dramas like this occur in a science, it is thought to be a science revolution and is considered a major event of a science's history.
- Model. A "model" is a vision, scheme, or representation of reality. It is a construct to compare a phenomenon or explore a phenomenon in some way. Models are constructed over most anything in Nature. Examples are a globe for the Earth, a double helix model for DNA, and so on. Models play an important role in science.
- Falsifiability. Theories are "accepted" by scientists because of how well they "explain" a phenomenon. However, theories are only accepted as long as no findings "contradict" or oppose its teachings. Theories are targets for testing and criticism always as its possible they could be "wrong" or overthrown later

on. Scientists may overly believe in a theory as true, but later ages of scientists may discover something challenging a theory's belief.
- Beware mistake. Science is a human activity and mistakes are sure to happen. A scientist should check and recheck findings to look for mistakes in analysis. It is because science is a human activity that mistake is always present. Mistakes routinely occur in science and must be guarded against.
- Comparison and contrast. This refers to comparing a phenomenon to something else and then commenting on similarities (same qualities) and differences. An example is to compare a cloud to a pillow or water with lava.
- Make predictions. If a theory does a great job of explaining phenomena, it can then go on to predict something new about a phenomenon. This usually results in science discovery and leads to many a historic moment within a science.
- Account for all the data. A good theory should "explain" all the data or account for all the phenomena it discusses.
- Always doubt. Some theories tend to be so "powerful" that they do not ever seem to be wrong. However, if a scientist believes in a theory too faithfully, he can be lead astray into believing that it always right. Theories are accepted bodies of beliefs until evidence comes along to show it is wrong. A theory can stay this way for ages and it is a hard task to overthrow a great theory in science.
- Facts prove themselves. If a natural feature is said to be "reproducible" or demonstrated time and again in experiment, it is thought a "truth" of Nature has been found. Truths of Nature are the following: light exists, the Sun does shine, the Earth spins, the speed of light is a constant, mass and energy exist, and so on. Identifying something as existing is to say it is a fact.
- Professionalism. The scientist must use caution, rationality, refinement, and respectability in the activity of science. He must not expose himself as an amateur, wreckless character, or abuser

of science. He must uphold the honor of science and demonstrate his competency in science.
- Publish facts in journals. Professional scientists when they discover something new in a science present their new knowledge in magazines called journals. A journal represents a forum whereby scientists can learn about the "goings-on" in a science and learn about new knowledge.
- Lecture and teach. It is the responsibility of scientists to communicate knowledge so science students can explore science on their own.
- Press conference. If a science finding is so extraordinary, then scientists may organize a press event to publicize it.
- Thought experiment. Thought experiments are dramas where scientists think about a phenomenon. They can be useful in exploring and discovering new things.
- Creation of a paradigm. A paradigm (para "dime") is a principle or central belief a theory, a "science age", or body of beliefs may have. Paradigms are created to "explain" a theory and elaborate on its meaning. A paradigm can be magic, the "Newtonian clockwork", the principle of relativity, science-like attitudes, and so on.
- Use reason. Phenomena can be investigated with intelligence and frequently can be reasoned out in their behavior. A phenomenon that routinely follows a rule is said to obey a natural law. Natural laws are fundamental to a science as laws comprise the basic teachings of a science. Overall, this is the act of trying to impose order, to comprehend the phenomena, and elucidate a theory of what is going on.
- Deduction. This is the act of trying to reason something out of a discovery, issue, or fact in science.
- Use mathematics. Mathematics is the science of numbers. It is an ancient subject and has proven itself time and again as being useful in science. It is used often in science to express natural laws and is found in chemistry and physics.

THE GRAND SCIENCE OF PHYSICS

- Beauty in explanation. If a theory has a simple, basic belief and shows its power in explaining natural things, it is said to be "beautiful" in some sense.
- Scientific proof refers to a "positive result" from an experiment that says something exists or is real.
- Scientific law. A scientific law is a rule by which natural things obey. Science knows many kinds of law and they are deeply studied and technical to discuss.
- Objectivity. Objectivity refers to viewing things for just their facts. It refers to not being swayed by emotion, cloudiness, ridicule, criticism, and appeals to feelings or sentiments not involved with the scientific process.
- Peer review. This is the activity of having fellow scientists review findings, conducting experiments, and commenting on an act of science. The scientist should not be prideful over his alleged competence or worth, he should expect and demand that his colleagues will probe him, doubt him, and question him on his findings. He should expect this and accept this as mistake can be guarded against and his worth as a scientist proven.
- Be open to suggestion and new ideas. Science thrives when its thinkers can freely share knowledge, ideas, and findings among each other.
- Use the laboratory. A laboratory is a scientific place to handle specimens, to probe, and to do research. Many a science has its corresponding laboratory. The science professional understand what a laboratory is and how to use one.
- Expeditions. Sometimes a scientist in doing science needs to travel to some place. For this he will organize an expedition to travel somewhere.
- Obtain funding. Sometimes a scientist is swayed by curiosity, he then may seek funding for an expedition to explore some science issue. Sometimes science projects prove so enormous,

SCIENCE

technical, and expensive, then the backing of a science organization is needed to fund an expedition or drama.
- See possibility. This is the activity of probing some thing and then seeing something possible out of its issue like a technology.
- Consult resources. Sometimes the scientist must read widely and deeply to research some issue in science.
- Get help. Sometimes science is just so complex to do, a scientist may resort to asking colleagues, assistants, and staff to help in research.
- Use technology. Science is a complex activity and often the scientist must use computers, lab equipment, and tools to do science.
- Avoid sensationalism. Scientists tend to be calm and level-headed characters. For this reason, they do not make wild claims like perpetual motion, revolutionary statements, and grand statements. However if a discovery or issue is incredible (like the atom bomb), it will naturally acquire its share of sensationalist statements and drama.
- Speculation. This refers to acts like trying to predict the future, discuss possibilities, and looking into the unknown. It is done by thinkers trying to extend science or see into a science what is presently unknown. It is an activity mainly practiced by science fiction writers, science visionaries, and science dreamers.
- Oppose gullibility. Scientists should not "rush" to believe in a claim should an experiment show it is "right". Claims need to be tested time and again to make "absolutely certain" that what is said to be real is in fact reality.
- Use heresy cautiously. Sometimes scientists will discover a fact that can overthrow a theory. Such dramas are not often in science history, but can be revolutionary. The scientist if he finds a fact that can overthrow a theory should be highly cautious in what he claims.
- Know history. History refers to what went on before in a science. A scientist should know what others had done in the past

for science as he follows in the footsteps of those who came before.
- Analysis and interpretation. This refers to extensively thinking about science findings to identify mistake, cloudiness, exaggeration, or other problems of thinking.
- Nomenclature. This refers to using the time honored traditions and methods to say that something has been discovered or how science is done.
- Jargon refers to the words (terminology) used by scientists to discuss science issues, results, and other topics.
- Credit is the activity of commending scientists who made a discovery.
- Orthodox. When a theory gains an immense amount of evidence, it is said to be believed in. Belief in a theory can become so powerful that it is now regarded as an "orthodox" teaching (or of the establishment). Theories like Newtonian mechanics, quantum theory, and relativity are considered orthodox theories.
- Work within the established order. In the modern world, science is now so advanced and complicated that the scientist often needs the support of a government, school, college, institute, or some powerful entity to finance and support scientific endeavor. A scientist should not be seen as a rogue or revolutionary as support may desert him and he winds up disgraced and denounced within his own science profession.
- Patent and technology. If a scientific finding proves to inspire a technology, a scientist can apply for a patent describing what he has discovered, the type of invention it is, and claiming ownership over the idea or innovation.
- Serve science. Science is a grand cause, a consequence of civilization, and an endeavor of Man. Uphold its honor and act with integrity.

In all, the scientific method is a time-honored practice used throughout the sciences. It had its origins with men named Biruni, Ibn

SCIENCE

al Haytham, Aristotle, Greek philosophers, Francis Bacon, Copernicus, Galileo, Newton, Kepler, and many others. It is a method that lets scientists investigate natural phenomena and lets them understand just what they are studying. The scientific method is taught worldwide and the reader is encouraged to explore this most fascinating subject in more serious works than this.

Good vs. Bad Science

Good science refers to thoroughly using the scientific method to explore a phenomenon. It means to exhaustively explore, research, study, and elucidate the nature of a phenomenon. Good science refers to dramas that win Nobel prizes or gain acclaim for competency in how an investigation is done. Bad science refers to shoddy, incompetent, and reckless behavior in doing science. Bad science involves making wild claims, not using the scientific method, not using peer review, and using bad methods in an analysis. Scientists generally review each others work to guard against bad science and to encourage that good science is done.

An Analogy of Science

Science is about exploring natural qualities, learning knowledge, and composing theories. An analogy to better explain the activity of science is the following:

When you look in the sky, you can see clouds passing away in the breeze.

The act of seeing clouds in the sky is called "observing" clouds. A person who sees clouds is then an "observer". Sensing most anything is an act of observation and anyone is an observer.

The person observing clouds would see they are far away in the sky, are colored white, appear fluffy or scattered, and "blow" on by as they "fly" in the sky. All these are observations on the nature of clouds.

Suppose someone makes a claim about clouds like the following: clouds are water vapor. Clouds imply storms are coming. Clouds are forever.

These are statements that can be believed true or false.

If you believe without any evidence, this is said to be an act of trust or foolishness. However, this is not an act of science.

A science-minded thinker would criticize and doubt these statements saying all this is nonsense. He would demand "proof" or a "test to confirm these statements are true". This is where an experiment is performed.

Experiments are "tests" to determine if a claim is true or false. They can be simple or complicated dramas and science knows many epic dramas.

The claim that "clouds are water vapor" could be found to be "true" if it rains or if it is found clouds can become rain.

The claim that "clouds imply a storm comes" can also be tested if a thunderstorm or tornado comes. If no storm comes then it is thought this claim is false.

The claim that "clouds are forever" can also be tested. An act of observation to see that they move and disappear "proves" this claim is wrong as clouds are not eternal.

A theory about clouds would go like this based on the following observations and experiments. A cloud is a "mass of water vapor" that is not forever, can become rain, and may or may not imply storms come. This theory is a body of beliefs about the nature of clouds. Its main principle is that "clouds are water vapor".

Many observers may come and doubt and criticize about the nature of clouds. This is only to be expected. However as each and every test about the nature of clouds implies that clouds are water vapor, the "theory" that clouds are water vapor grows in confidence. Eventually, this confidence grows so strong that theory faith appears and scientists believe no other way. So the theory that clouds are water vapor grows evermore believable and scientists in time believe no other way.

In time a fact may appear that clouds may not be water vapor. If that happens and it is not likely, this can create a crisis whereby scientists are thrown into confusion over the nature of clouds. This could start a drama whereby new thinkers propose clouds are something else like dust or soot. This would have the effect of "overthrowing" the

SCIENCE

"established" theory and replacing it with a new theory (clouds are dust). This would be a "revolution" in theories about clouds.

In all this demonstrates how scientists think and how science operates.

The Advancement of Science

Throughout the ages, there were thinkers who would perform an experiment, discover a phenomenon, explore a phenomenon, discover a natural law, or invent a technology. These people would go on to advance science by their acts. Science advances in leaps and small jumps. Most scientists are content contributing small and insignificant facts to a science while others from time to time do something monumental. In all science can advance as slow as a snail or as fast as a cheetah. It depends on the circumstances and the people who make science advance.

The History of Science

Science is an ancient subject going back ages. It is even now always changing with new facts and thinkers appearing all the time. History is a record of what has gone on in the past and represents the trends of how a science advances. Epic thinkers who have made great advances in the understanding of science are the following (please explore their lives and contributions):

- Greek philosophers. These thinkers of ancient Greece were among the world's first proto-scientific thinkers who shaped later ages of thinking. Famous names here are Aristotle, Socrates, Archimedes, Pythagoras, Euclid, Democritus of Abdera, Zeno of Elea, Thales, Aristarchus, Hipparchus, Hippocrates, Herodotus, Thucydides, and many others.
- Roman thinkers. Ancient Romans are known for building aqueducts, writing works, devising tools, and other acts that would find their way into science.
- Lucretius is a Roman thinker who published an epical book describing Roman views of Nature and in general the Roman worldview.

- Ancient Chinese thinkers would write works, invent the compass and the rocket, stargaze, and discuss ideas that would be used in modern science.
- Ancient Indian philosophers discussed ideas like atoms, stargazed, wrote works, and discoursed on ideas today found in modern science and philosophy.
- Ancient thinkers like Herodotus, Manetho, tribal shamans, court priests, scribes, and the like would keep records of what has gone on before. In some sense this began the subject of history.
- Ancient thinkers like Galen, Hippocrates, tribal shamans, court healers, medicine men, and the like would take to trying to cure people of diseases. They would use spells, potions, invoke the gods, pray to God, chant, and perform ceremonies to try to heal. Often times, the sick were told to seek the help of a wizard or a witch, bathe in "healing waters", perform spells, or take potions to "cure" them from disease. In some sense medicine had its origin in these practices.
- Babylonians and Egyptians. Thinkers from these societies would compose the modern calendar, kinds of clock, tools used in engineering, and other devices.
- Alhacen. He was an Islamic scientist from the Golden Age of Islamic Science. He and others flourished in the Middle Ages at the House of Wisdom in Baghdad, Iraq.
- Native Americans. When Columbus discovered America, he encountered various tribes of indigenous Native Americans. These peoples were known to consume foods never before seen, use primitive tools, and used kinds of technology the Europeans did not have. Some of their tools have found their way into science.
- Ptolemy is an ancient thinker who would write the epic work, the Almagest. His thinking would shape "science understanding" in the Middle Ages.
- Gutenberg would invent the printing press thus leading to the modern era of books and literature. This was a crucial step

SCIENCE

leading to a more literate era and helped create the modern science age.
- Ancient alchemists would tinker with natural substances like rocks, minerals, acids, explosives, and the like and in a way "accidentally" invent chemistry.
- Ancient herbalists (collectors of plants) would investigate various native plants and animals and in time "accidentally" invent sciences called botany and biology.
- Various ancient thinkers would domesticate wild animals leading to pets, farm animals, beasts of burden, and zoo animals.
- Ancient herbalists, shamans, tribesmen, and many others would take to cultivating plants for food, medicine, and other uses. In some sense, agriculture began here and would continue on to become factory farms and the Green Revolution.
- Primitive tribal shamans would collect plants, learn about animals, investigate natural substances, stargaze, observe the weather, and do acts that would lead to the "accidental" founding of sciences called chemistry, biology, astronomy, meteorology, and so on.
- Ancient storytellers would take to inventing stories of how the universe and the world were created. They would embellish their stories with heroic tales of heroes, the gods, demons, spirits, and acts of magic. In some sense, the science of cosmology had its origin in the discussion of creation myths.
- Thomas Kuhn and Karl Popper. These are men who left epic works describing the nature of science and are considered authorities even now.
- Renaissance thinkers. This collectively refers to men who shaped science centuries before the 20th century happened. Celebrated names here are Kepler, Copernicus, Newton, Galileo, Descartes, Liebniz, Bacon, Kant, Dalton, Boscovich, Huygens, and many others. Their works are taught routinely in philosophy courses.

- Natural philosophers. These are people who took to asking science-like questions of Nature before science was invented. Many such thinkers like this lived in ancient times, the Renaissance, the Enlightenment, the Middle Ages, etc.
- 18th century thinkers. This refers to prominent thinkers of science who lived in this time. A famous thinker of this era named Benjamin Franklin would perform a kite experiment proving lightning was electricity as well as being a major figure in the American Revolutionary War and Constitutional Convention.
- Dmitry Mendeleev, a Russian chemist would write the Periodic Table of the Elements thus beginning the modern era of chemistry.
- Darwin would conceive evolution and begin a controversy that would shape modern biology and religious discussion.
- Gregor Mendel would conduct experiments that would lead to genetics.
- Thinkers like Pavlov, Freud, Adler, Jung, and many others would found the modern science of psychology.
- Thinkers like JB Rhine, Zener, and many others would explore the paranormal and attempt to found a science on it.
- Thinkers like Alfred Nobel, Gatling, Browning, Napoleon, generals, and others would make advances in the understanding of warfare. This would result in the invention of new and more powerful weapons leading to highly destructive wars.
- 19th century thinkers. In this time, people like Darwin, Freud, Faraday, Oersted, Maxwell, Kirchoff, Mach, and many others lived shaping modern science to come.
- Edison and Tesla are inventors and contemporaries who would create monumental inventions in electricity and other domains. Between the two of them, the modern world was founded by their inventions into electricity.
- Various inventors have conceived of devices that have influenced civilization. There have lived many "pioneers" who

SCIENCE

discovered something new, created a technology, or did something original to influence science in some way.
- Thinkers like Cousteau, Ballard, Rickover, and many others would pioneer deep sea exploration, aquatic technology, and many innovations.
- Alfred Wegener would study a map of the Earth and notice that the continents seem to fit like a giant jigsaw puzzle, this would lead to the discovery of plate tectonics.
- 20th century thinkers. In this time, there lived many epical scientists like Einstein, Bohr, Pauling, Bardeen, Hawking, Rutherford, Curie, Rabi, Oppenheimer, and many others who shaped this era of science, sometimes called the Golden Age of Science.
- Robert Goddard would experiment with rockets leading to the Space Age.
- Werner Von Braun would take an interest in rockets in Nazi Germany and go on to a career in NASA leading to the Moon Landing.
- Thinkers like Fermi, Szilard, Einstein, Hahn, Strassman, and Meitner would realize that the atomic nucleus can be split leading to the atomic bomb and nuclear power.
- Henry Ford, Benz, Diesel, and many others would build engines and later automobiles beginning the modern era of cars, busses, and other device.
- The Wright Bros. would build the first airplane and begin the Age of Flight. The Montgolfier Bros. and Zeppelin would pioneer balloons and blimps.
- Marconi, Hertz, and others would pioneer radio.
- Charles Babbage, Ada, Hollerith, and many others would explore making machines that could count thus leading to the invention of computers.
- Alexander Graham Bell would invent the telephone.

- Alfred Nobel would invent dynamite and found science awards called the Nobel Prize, destined to become science's ultimate award for achievement.
- Philosophers discuss ideas like will, force, meaning, change, esthetics, and many other ideas. They have contributed to science in immense ways. Famous names here are Godel, Whitehead, Aristotle, Kant, Descartes, Liebniz, and many other names.
- Thinkers like Newton, Liebniz, Euler, Descartes, Gauss, Riemann, and many others would shape mathematics and hence contribute immensely to science.
- Explorers like Columbus, Magellan, Cartier, Drake, De Soto, Da Gama, Balboa, Cortes, Pizarro, Lewis and Clark, Peary, Amundsen, Cook, and many others would go on voyages, make discoveries, and contribute something epical to the creation of the modern world of science.
- "Curious thinkers" refers to people who would "explore" the paranormal, the occult, UFOs, and other exotic subjects in an attempt to make them subjects where science can be used to solve their mysteries.
- Lawrence would experiment with a machine called the cyclotron and inspire modern particle physics.
- Hubble would discover the expansion of the universe.
- People like Herschel, Adams, Galle, LeVerrier, Piazzi, Tombaugh, and others would discover comets, asteroids, and planets.
- Curie, Becquerel, Roentgen, and many others would explore X rays and radioactivity.
- Thinkers like Witten, Schwarz, Bell, Bohm, Penrose, and many authors would conceive theories and make discoveries that would lead to the modern age of physics.

There are in reality many epic stories from the ages when a great advance in science took place. The people and these dramas of lore would shape science and become science mythology. Please explore their stories in the many textbooks on science.

SCIENCE

Protoscience

This word refers to subjects that are beginning or changing to become a science. As of now, they are not regarded as sciences as they seem too suspicious, primitive, or uncertain of their findings. An example is parapsychology. It is overall the study of such issues of the paranormal as telepathy, prophecy, bilocation, and so on. These issues have been discussed for centuries, yet there has been no proof that anything has ever occurred. It is thought that maybe in the future, some of these issues will be found to be real and hence can take "their place" as being regarded as phenomena. When this happens, it is thought this subject will finally be recognized as a science. However for now, this subject is considered just too controversial to be awarded the respectable title of a science.

The Metric System (Weights and Measures)

There is in science a system of "units" by which "quantities" are measured and valued. Examples of such units are the meter (length), liter (volume), coulomb (electric charge), newton (force), kilogram (mass), second (time), and so on. Devised in France, it is a system of units used worldwide and also in science. Virtually every "quantity" in science has its corresponding metric unit (also called SI unit). An education in units is vital to learning science. Also many countries maintain their own system of weights and measures (units) and examples are the British system, cgi, and many others.

Please read about the nature of the metric system and other systems of weights and measure.

The Technology of Science

Throughout the ages, many scientists have invented tools by which science can be practiced. Many of these tools have gone on to be regarded as classic inventions. Today many of these tools are being enhanced to become ever better kinds of instrument in the activity of science. A list of tools known to science are the following:

- telescope –centrifuge –particle accelerator –space probe –chemistry equipment –thermometer –weather devices –deep sea submersibles –probes –balances –binoculars –microscope –electron microscope –flasks –pressure chambers –lasers –computers -calculators -wind tunnels -test sites -diving equipment

Please read about the many tools of science.

The World of Scientific Research

Around the world at laboratories, colleges, institutes, and other places, scientists are hard at work trying to explore the unknown of the sciences. Experiments are going to test the limits and boundaries of science to discover the unknown and to solve its mysteries. Today's age of science research is a fabulous era where many epic stories are taking place as of now. Today's era of the 20th and 21st centuries is considered a particularly exciting time for science. In this era, relativity, quantum mechanics, lasers, computers, new astronomical objects, and new effects have been found. This age continues on with even more sensational dramas arising whenever.

The Scientific Community

There is a community of thinkers (called academics) that make up a scientific community. They are professors, lecturers, researchers, and thinkers who maintain devotion to a science. They act as a community of thinkers who research a science, write textbooks, teach a science, discuss a science, maintain its cause in some way, and act as "guardians" of its teachings and methods. They organize themselves into associations or societies for the purpose of discussing a science. Scientific communities usually consist of professors and students gain "admission" to its community by obtaining academic qualifications called "science degrees" (usually doctorates or some college degree). Academics (scientists) in a science generally go on to become teachers or officials of the science in some way. Some scientists go on to do epic moments in a science (perform a key experiment, compose a theory, or make a discovery) and become pioneers of a science. Some scientists win

SCIENCE

acclaim for their acts of science and become historical figures in a science. Others may win the Nobel Prize or some other award. Eminent associations of scientists are examples like the Royal Society or the US National Academy of Sciences. Professors of rank and accomplishment are called "professor emeritus" and there are many classes of professor like associate, visiting, assistant, and full types. A scientific community is then a collection of academics that comprise the "professional world" of a science.

Science Professor, Scientist

Various students will go to college, "major" in a science, and grow into being scientists. They may attain master's or "doctorate" degrees and attain to a university position called a "professor". A professor is an academic who researches a science, teaches, mentors graduate students, and in some way "professes" expertise in a science. Many colleges maintain staffs of them and professors make up the "professional" community of a science.

Science Paper (Academic Paper)

A science paper is a document that discusses research, an act, an issue, or a problem in science. It is usually authored by a lone scientist or a team of them. It generally is authored by a senior scientist working in cooperation (collaboration) with graduate students or colleagues (equals in science). Science papers are generally contributions to a science's learning and can be momentous dramas in the history of science. Science papers are the output of science work.

Science Journal

A science journal is a book, magazine, or document where science papers are published. It is a forum to discuss science issues, research, and make contributions to science. A science journal is published by a respected organization in science, overseen by editors and staff, and involves rigorous guidelines before a science paper can be published. Science journals are professional forums in a science. Famous science journals are the Astrophysical Journal, the Physical Review, and others.

Science Department

In many colleges, there will be a building that houses departments dedicated to some science. Such sciences can be geology, geography, physics, computer science, biology, psychology, linguistics, or something else. Science departments will have a staff of professors overseen by a chairman. It will have offices, lecture rooms, lab rooms, and resource rooms for a science. It is here that a science is taught on campus and students are trained in a science.

Laboratories

These are special and unique places where science can be done. They are usually found in colleges, institutes, government or military sites, or wherever. They are staffed by professional scientists complete with the tools of the science. Here in "labs" epic discoveries in science are made. Famous labs are the Cavendish laboratory, the Niels Bohr Institute, Fermilab, CERN, the Kurchatov Institute in Russia, and so on. Labs are known to have their dangers and should only be handled by professional scientists and staff assistants. Laboratories come in many kinds like the following:

- chemical laboratory -particle accelerator -observatory -neutrino telescope -physics laboratory

Science Institute

When a science grows immense in its knowledge and has many issues to study, an organization called a "science institute" is founded. It is essentially a "place" often as a branch of a college where science academics can research a science. Here many epic breakthroughs in a science can occur and new technologies often arise. Science institutes tend to have a director (ranking professor), researchers, professors, laboratories, and receive funding from a college, government, or other organization. Various kinds of science institute are observatories,

SCIENCE

oceanographic institutes, weather research stations, particle physics laboratories, space institutes, and so on.

Research and Development (R&D)

R&D is a practice in science of taking science teaching and turning it into technology or exploring subject matter. It is practiced in private industry, national laboratories, science institutes, colleges, and other places of professional science. People who attain to the doctorate degree may graduate and then go on to do 'post doctoral research'. Many a researcher goes on to attain the rank of professor.

Science Literature

Scientists like to publish or make public their knowledge about a science. This is done to communicate to other scientists and the world the knowledge that they know. Epic works where science knowledge appears are textbooks, pamphlets, journals, websites, almanacs, encyclopedias, popular science writings, magazines, and so on. It is in these publications that science can be told to a wider audience in the belief a science can advance if others know about it.

Science Library

Here in libraries usually found in colleges, institutes, and elsewhere, collections of books are maintained on the many issues of science. Here collections of science magazines, books, periodicals, videos, audio books, and so on are maintained. Books by leading authors, textbooks, interest books, and other literature are stored here. For those who have an interest in science, a science library is a place to visit. Famous "libraries" known to science history are the Bodleian, Library of Congress, Library of Alexandria, US Patent Office, and so on.

Science Museum

A science museum is a place where science, moments of science, technology, and other things are put on display. They are kinds of "temple" to science glorifying or teaching science to youth and other people. Here science is displayed for youth to know the history of science and the people who made science happen. Famous kinds and institutes of science museum are:

- planetariums –physics museums –observatories –Lowell Observatory –Smithsonian Institution –Museum of Science and Industry –aviation museums –IMAX theatres of science –museums of science institutes –geologic museums –dinosaur museums –Library of Alexandria –Bodleian library –Baghdad Museum

Science Fair

This is a grand event where students of science or scientists present findings, display exhibitions, and give lectures on science issues. These events may involve competitions over science achievement, may involve the awarding of science prizes, and are dramas where the public can meet scientists.

Science Club

Many schools ranging from elementary all the way to graduate level may have science clubs. They are groups of students and teacher (faculty) who meet to discuss science or a science. They are fun groups to socialize, make friends, and go on science trips. Students are encouraged to join these groups. Students may go on to compete in science competitions arising from science clubs. Kinds of science club are the physics, chemistry, math, and biology ones. Kinds of science competition are Science Olympiad, Science Bowl, and more.

Science Patrons

Patrons are people and organizations that support, finance, lend credibility, or in some other way promote science. They may give money, found institutes, finance expeditions, or do acts to support science in some way. Acts of patronage are:

- awarding of science prizes –bestowing fellowships –award "grants" of money –support science initiatives –found an institute –recognize scientists –establish chairs -conferring the Nobel Prize -giving titles -promotions -endowments -scholarships -establishing foundations

Science Education

Science has in some way been taught for thousands of years. While it only began about the time of Galileo, science-like teachings have been taught since ancient Greece, Rome, ancient China, and many ancient societies existed. It is usually taught in schools and from there goes on to college and graduate school. It is usually taught to kids who then choose if they like science or not. From these people come the various ages of science and the people who make the history of science.

Science Literacy

This refers to trying to teach, publicize, and communicate information about a science. It is known that the general public is largely ignorant of science. Many efforts have been made to teach science, publish journals and newsletters, make TV shows, and do other acts to get science told to the public. It is considered a major challenge to get kids interested in science and to learn science.

Careers in Science

Learning science in college can lead to many careers. While it may not necessarily lead to a career as a scientist, science can open a lot of doors. People who trained in science in college can go on to careers like the following:
- astronomer -cosmologist -science writer -camp counselor -science official -professor -inventor -game designer -museum curator -explorer -diver -astronaut -college president -politician -patent examiner -pioneer -sailor -musician -philosopher -teacher -lawyer -science personality -TV host

Graduate School

This is a kind of higher education that comes after college. Here students will refine their science education and become "professionals" of a science. Many sciences gain their next age of professors and leading thinkers from graduate schools. Its in graduate schools that many epic moments of science are done and leading professors can mentor the next age of tomorrow's thinkers in science. Here students can gain degrees termed the master's degree or doctorate. After obtaining them, a student can go on to be a post-doctoral researcher and advance to the rank of professor. Going to graduate school tends to be expensive, requires having high grades, and passing a test called the GRE. It is for those considering a serious career in science.

Organizations of Science

Science began as superstitious activities that became refined by reason, testing, criticism, doubt, and research. It lead various peoples to question reality and seek out what was real or not. In time, there would grow up organizations that would try to represent science, champion science, or organize a science into a respectable activity. Presented here a list of organizations that in some way have championed the cause of science as a profession, "crusade", or activity "sacred" to civilization:

- Royal Society –American Association for the Advancement of Science –IUPAC –IUPAP –International Astronomical Union –National Geographic Society –National Science Foundation –American Academy of Sciences –Smithsonian Institution –eminent colleges –NASA –ESA –CERN –Roskosmos – Pontifical Academy of Sciences –MIT -JAXA -Committee for Skeptical Inquiry -space agencies -science foundations -Nobel Institute -weights and measures institutes

There are in reality many thousands of such groups. They are staffed by professors, volunteers, ranking academics, and eminent personages in some capacity.

SCIENCE

Academy of Sciences

Many national governments maintain an elite body of academics known as an 'academy of sciences'. Here professors who obtain eminence in a science may be 'elected' to join an academy of sciences. They are chosen for their position, rank, eminence, achievements, and science awards won. In such a group they may serve on boards, judge science requests, and serve science in some way. Being chosen for membership is a high achievement in science.

Science Awards

Science is a drama that has "blessed" and "cursed" Man in many ways. It is a grand drama to explore the unknown, discover something new, and find new technologies. Various organizations have founded "awards" that have tried to "glorify" or "commend" scientists who have done something epical for science. They usually consist of cash endowments, medals, citations, and honorifics to celebrate scientists and science itself. Presented here is a list of some of science's more famous awards:

- Nobel Prize –Rossi Prize –Kilby awards –MacArthur Genius awards –fellowships –National Medal of Science or Technology –Vetlesen Prize –Wolf Prize –Queen's citations –royal awards – commendations –statues –Planck Medal -Crafoord Prize -Abel Prize -Dannie Heineman Prize

There are in reality many kinds of prize and award and they can be explored further.

Glory of Science

When scientists have done outstanding work or contributed immensely to a science, they may become celebrities in their own lifetimes or in history. They may gain "prestige" or considerable influence in science to direct research, hold eminent positions, or be a voice of reason and civilization in the world.

Nobel Prize

In the 19th century, an inventor of dynamite named Alfred Nobel amassed a fortune in explosives. He dedicated his wealth to the creation of prizes to reward scientists and thinkers who have contributed immensely to the welfare of Man and the world. He established five domains to commend such people in physics, chemistry, literature, world peace, and medicine (later an economics award would be added). Each year since 1901, various people from around the world would be chosen by committees to receive the Nobel Prize for some contribution to science or to Man. They would be given cash awards, medals, and diplomas saying what they had achieved for the good of the world. Since then, the Nobel prizes have gone on to become science's most famous award. It rewards otherwise obscure people with fame and money for their epical achievements. Today, the Nobel prize tradition continues each year with grand celebrations held in Stockholm, Sweden. Winners of the Nobel Prize and also called 'Nobel laureates'.

Personality of Science

Some scientists may grow in science to attain the rank of professor. They may star in a TV show teaching science, advise on science, participate in a scientific drama, win the Nobel Prize, or do something that gains fame in science. From then on they gain a celebrity that makes them famous and a star in science. They may be invited to give speeches, to mentor students, to advise others, to work on projects, or just to lend their credibility to an endeavor. People who would gain fame as personalities of science are Newton, Einstein, Carl Sagan, Steven Hawking, and many others.

Science Disgrace

Scientists are held to high esteem in society for rationality, intelligence, and achievement. For this they should not embarrass themselves or discredit themselves by wreckless acts. Such acts include crime, boastful and unproven statements, wild speculation, neurotic behavior, insanity, lying in science, falsifying data, or taking credit for false achievements. When the scientist is discovered doing immoral or illegal acts, he can find his reputation destroyed in science. He may be dismissed from his position, have his titles revoked, lose his sponsors, and be shunned from professional science. Too many scientists suffer this fate when they do an act leading to science disgrace.

Scientific Revolution

There are times in the history of sciences when cherished beliefs seem to be and are wrong. Often times, events will occur that throw into doubt some kind of scientific belief. When this happens, thinkers arise to question findings, data, and theories. They may propose new ideas or overthrow existing theories. When this happens it is said a "scientific revolution" is occurring. These dramas go on to be sensational in science leading to many epic stories. While these dramas are rare, they do happen and may happen again.

Please read about these and other dramas when sciences have been forced to change their beliefs.

The Philosophy of Science

This refers to a branch of philosophy concerned with the nature of science. It is an age-old subject that compliments science teaching. Famous names here are Thomas Kuhn, Karl Popper, Descartes, Galileo, Aristotle, Newton, and many others. The reader is encouraged to explore this obscure and highly abstract subject.

The Future of Science

What is it that will science discover and discuss in the centuries to come? No one really can say. However its thought such issues may consist of suspended animation, cures for disease, alien contact, super-technologies, the grand unified theory, and quantum gravity. For now, this subject is a hobby for speculators, dreamers, and science writers trying to imagine the future of science.

THE GRAND SCIENCE OF PHYSICS

Blessings of Science

Science advancement has "blessed" humanity in many ways. It has given him labor-saving appliances, airplanes, boats, cars, nuclear power, life enhancing medicines, and new technologies like TVs, DVD, video games, and so on. Science has allowed Man to advance out of the 19th century into a more complicated time.

The Tragedies of Science

With science advancement, thinkers would come along and find ways to "misuse" its gifts for crime, war, terrorism, destruction, and carnage. Science advancement has allowed war to become vastly more destructive than in previous ages. In the 20th century, new weapons like flamethrowers, tanks, machine guns, atom bombs, nerve gas, bioweapons, chemical weapons, and others appeared. They were used in horrific wars called the world wars, Vietnam, Korea, the Gulf War, Afghanistan, and so on. Science in some sense allowed Man to become more destructive than he previously was in other ages.

Science Fiction, Futurology, Futurism

This is a branch of literature dealing with exploring the "future" of science. It explores science issues that today are only "speculations". It will discuss stories, have "heroes", explore "worlds", and go on adventures to explore what the future of science may be. Famous issues in SF are aliens, suspended animation, star travel, time travel, free energy, ultimate weapons, miracles, the paranormal, robots, and so on. Famous authors in SF are Asimov, Clarke, Wells, Verne, Dick, Lucas, and modern authors. Futurology is a science that studies the future in trends and beliefs. Futurism is the practice of predicting something for the future.

Technology & Invention

Science has given the world many "gifts" that former ages did not have. These gifts have become tools found throughout society

impacting virtually all domains of civilization. When science beliefs or findings are made into tools, it becomes technology (or inventions) used by scientists and others. Technology has greatly impacted the world and the activity of making newer kinds of technology is called inventing. Famous kinds of technology that resulted as "gifts of science" are:
- cars –airplanes –rockets –blimps –spaceships –missiles –atom bombs –electrical devices –magnetic devices –computers –radio –radar –telescopes –microscopes –clothing –shoes –boats –jeeps –tanks –guns –ammunition –weapons –knives –axes –fire –wheel –simple machines –chariot –buildings –medical equipment –toys –games –books –brick –concrete –steel –alloys –nuclear power –lighting –fences –farm equipment –video game

There are in reality many kinds of technology all of which impact the world, civilization, and lives all around the world.

Inventing

Inventing is an offshoot of science to conceive new ideas, innovate, make new kinds of technology, combine sciences or issues, and so on. Inventing has lead to things like the telephone, computer, video game, atom bomb, and so much more. Famous names here are Edison, Tesla, Archimedes, Nakamatsu, Baer, Bushnell, and so on.

Popular Science

This refers to all kinds of machine, teaching, book, and business that involve the general public with science. It can refer to science competitions, science stores, magazines that sell science tools, science kits, and much more. It is a hobby and subject for the curious hobbyist scientist than anyone else.

Science Politics

Science has grown so immense it now affects the discussion of politics around the world. Today, science issues affect the world in such topics as global warming, the ozone hole, the world energy crisis, pollution, whaling, and so on.

Mysteries of Science

In ancient times, the world was a great unknown to Man. He went about exploring and investigating the many features of Nature and in this science would have its birth. Today, science has advanced far and given the world such things as atom bombs, electricity, magnetism, nuclear power, toys, rockets, and so on. Though science has solved many mysteries of Nature, there are today many issues that stubbornly refuse solution. Today, scientists are hard at work trying to solve these mysteries. It is thought the solving of these mysteries would be a grand moment of science and lead to discovery and dramas to be. Presently, science cannot solve the following mysteries:

- Cure for cancer –Grand unified theory –Quantum gravity – Search for alien life –Explain the Big Bang –Solve the mystery of the fate of the universe –Unify more forces –Explain consciousness –Cure AIDS and many diseases –Explain the origin of life -UFOs -Alien abductions -Ancient astronauts

Science is today filled with many mysteries and it awaits what the future will hold with these enigmas in science.

Overall, science is a body of learning and teaching that grew out of superstition, questioning, and natural philosophy. It is a movement to investigate Nature and probe its many secrets. It blends such subjects and issues as reason, logic, philosophy, religion, investigation, skepticism, and much more into a movement and practice that seeks to understand Nature. It had its formal beginning in the time of Galileo, Kepler, and Newton as learning evolved to separate science from superstition. It is today a world activity practiced at colleges, institutes, and places wherever. In the next chapter, we will explore a branch of science concerned with discovering Nature's fundamental laws. It explores such issues as mass, energy, space, time, concepts, atoms, change, and effects. It had its origins in ancient Greek philosophy and superstitions and has become the sensational and fantastic science of now. Today, it's

SCIENCE

a rich world of issues ranging from atom bombs to gravity waves, from relativity to black holes. It continues to be advanced and explored and no one really knows what is next for this science. We will now explore a discussion on physics.

> "Science itself is not religion. It is a cause and a way of thinking. It champions reason, it explores cautiously, it tries to explain away from superstition. Because it would prove its worth so often, then Man would adopt it as a way to see into Nature, to understand the complexity of things."
>
> -science discussion

Chapter 2
Physics

"Physics is a science that is like religion in cause and belief. Both want to understand Nature for its fundamental. Both seek to know the inner workings of Nature. Religion would invoke God, while physics invokes reason and physical law. In some sense both lead to the same ends. In some sense physics seeks to know if God exists and if He made Nature after all."

-physics discussion

Physics

The World of Physics (Physical Reality)

There is a branch of science that studies the nature of the physical universe. Called physics, it has been described as the most fundamental of sciences. It is in reality a vast labyrinth of related sciences all interrelated under the word physics. Physics as a science studies the properties of Nature like heat, light, electricity, gravity, mass, space, time, events, processes, effects, sound, radio, vibration, and many related subjects. It is today a complex subject taught in virtually all colleges and requires years of schooling to get a career in. Presented here is an overview of this very fascinating science.

Before Physics

Physics today is a mature science which is a very specialized profession. However, this science had its origins in ancient superstitions and occult practices. Ages ago, various thinkers took to asking questions about Nature like "Why do things fall down?" and "Why does the Sun shine?" Thinkers who posed such questions were called natural philosophers and they often acted as court physicians, astrologers, and as priests in ancient societies. They were among the first thinkers who asked questions of Nature and sought answers that did not involve magic or superstition. Physics in some sense had its origins in superstitious practices like astrology, alchemy, and the occult. Along with the writings of Greek philosophers, medieval Muslim thinkers, and other philosophers that physics thinking had a kind of origin. Its from these occult beginnings that physics took shape and would become the modern science of today.

Nature

Nature is a word that stands for universe, cosmos, and indeed all reality. It is the sum total of universe, Sun, Earth, planets, stars, Man, life, and all things of existence. Physics seeks to analyze Nature's many effects and qualities. It seeks to study and learn the secrets of natural qualities. Nature has been called on occassion by the following labels:
- The Cosmos -The Universe -Physical Reality -Reality -Everything -The 10,000 Things -The World -The Outside World -Creation -Physical Universe –The Void –God's Universe

Physis and Logos (Greek ideas)

These are ancient Greek ideas on reality, order, meaning, and substance. Their discussion began with men named Aristotle, Democritus, Thales, Socrates, and many others. In time, such words as physics, metaphysics, -ology, logic, physiology, and others would be invented. It is from these words and their discussion that modern physics arises.

What is Physics?

Physics is the study of physical reality in all its ways and its secrets. It employs basic ideas like space, time, mass, and energy. It would diversify into a vast labyrinth of sciences all collectively gathered under the word 'physics'.

Metaphysics

This subject is known from ancient Greece. It it the discussion into speculations on the nature of matter, fundamental substance, space,

PHYSICS

Nature, and many related issues. Aristotle, an ancient Greek philosopher is credited with being a major thinker in it as well as others. It has gone hand in hand with physics through time evolving and expanding of its discussions. An education in metaphysics enhances a physics education immensely. Please read about this and Aristotle's writings outside of this book.

The Occult, Superstition, The Magical

The occult is a collective term referring to practices like crystal ball gazing, prophecy, divination, ouija, casting spells, incantations, invoking spirits, praying to the gods, etc. Collectively, they are called superstitions and Man has been practicing them since his beginning. Today, they have largely died out with the onset of science, but still continue in primitive parts of the world. Occult practices in time would go on to "birth" many kinds of science and one of them is called physics.

Ancient Debates

Thinkers for ages have wanted to explain the "natural world". They took to discussing and questioning Nature and devised issues that would become "classic" in later physics and philosophy discussion. Presented here are a range of issues to explore and discuss in class:
- The nature of space –What is time –The nature of atomos (or atoms) –The nature of energy –The nature of mass –Discussion on ether –The nature of force (or forces) –The nature of substance –The nature of order and chaos –The nature of spirit and the gods –Discussions on ideas like physis, logos, apeiron (infinity), nous, mind, life, being, nonbeing

Physics Ideas (Concepts)

In time, various thinkers would invent ideas by which physics could be thought about. These ideas went on to become the basic beliefs of the science. Presented here is a list of ideas elementary to physics.

Mass. This is the idea of substance, stuff, and thing by which material objects are made of. Mass can also be called matter, koinomatter, and substance.

Energy. This is the idea of power, motion, and work by which objects move.

Massergy. This is the combined concept of both mass and energy.

Space. This is the idea of emptiness, shape, and form so vital in physics.

Time. This is the idea of change and motion as measured by a clock.

Spacetime. This is the combined idea of space and time.

Force. This is the idea of an influence through space or is felt in some other way that causes motion. Examples are centrifugal force, gravity, electric force, contact force, magnetic force, centripetal force, the strong nuclear force, and so on.

Interaction. Another name for force often preferred in physics by professional scientists. It can mean the contact of fields with each other or particles with each other. Various subatomic forces are also referred to as interactions.

Field. This is the idea of an energy presence in space able to cause forces. Examples are the electric field, magnetic field, and gravitational field.

Mind. This is a general word meaning all thoughts, imaginings, concepts, ideas, intuitions, visions, and constructs of thinking (or of Mind).

Inertia. This is the idea of the tendency of masses to resist changes in motion.

Motion. This is the idea that masses can change position in space with time. Famous ideas here are speed, acceleration, impulse, torque, spin, and momentum.

Worldview. This is the idea of a "vision" to view and know the universe or world. A worldview can be the Earth is flat, an infinite sea of space and time, and so on.

Work. This is the idea of something being done or power is used with energy.

PHYSICS

Power. This is the idea of being able to do something, it is related to energy.

Weight. This is the idea of a mass with an acceleration causing "heaviness".

Equation. This is a mathematic statement of order and behavior. Examples of equations are $E=hf$, $a^2 + b^2 = c^2$, $F=ma$, and so on. Physics uses many equations.

Heat. This is a word for a kind of energy able to cause motion with words like hot or cold, temperature, absolute zero, thermal energy, convection, insulation, and so on.

Light. This is a word for a kind of energy to be able to see and comes in types called colors. Colors can be expressed by the "word" ROY G BIV.

Electricity. This idea means electron flow.

Magnetism. This is an energy that consists of two "poles", lines of force, acts in iron and steel, has a magnetic field, and many other qualities.

Gravity. This is a word for a force that pulls down.

Quantum. This is a word that means "packet" or an amount of.

Atom. This is a very, very small entity consisting of a nucleus and electrons.

Nucleus. The center of an atom.

Proton. A particle of the nucleus with a positive charge.

Neutron. A particle of the nucleus with no charge.

Electron. A particle in orbit about the nucleus with a negative charge.

Wave. This is a motion of water or some continuum in an up and down movement. It can be expressed as a "sine wave". Throwing rocks into water demonstrates wave behavior.

Particle. This is a word that means an amount of mass, a sand grain is a particle.

Wavicle. This is an object that's a combined particle and wave.

Physical. This is a word that means of Nature, of the universe, of reality.

Physical law (natural law). This is a word that means a rule by which Nature obeys of its behavior. A physical law can be expressed as a mathematical equation. Examples are $E=mc^2$, $F=ma$, $E=hf$, and so on.

Event. An "occurrence" like a finger snap, eye blink, ball bounce, etc.
Universe. The sum total of all space, time, mass, and energy.
Effect. An "occurrence" involving mass and energy displays.
Speed (Velocity). This is the rate that a mass object moves.
Acceleration. This is the rate at which a speed changes.
Position (Location). This refers to place in space.

There are in fact many kinds of idea (or concept) each with their meanings. For a deeper appreciation for physics concepts, the reader is encouraged to explore more ideas and their meanings.

Phenomena ("What is reality after all?")

Physics is a deeply complicated science dedicated to studying various aspects of Nature. Nature is filled with vast amounts of effects called phenomena. Many require deep discussion and hence a detailed account of them is not appropriate here. However, physics does concern itself with the following effects routinely taught in schools everywhere. Examples are:

Mass. This refers to kinds of substance like water, air, ice, rock, sand, soil, plant matter, flesh, subatomic particles, gases, liquids, solids, and so on.
Energy. This refers to such natural qualities like heat, light, gravity, and so on.
Space. This refers to the "emptiness" in between stars and masses.
Time. This refers to the mysterious passing of event after event by clocks.
Forces. These are energy effects that cause motion, four fundamental forces are known to exist.
Fields. These are energy disturbances in space known to give rise to forces.
Light. This is an energy emitted by stars and allows vision and lasers.

PHYSICS

Electricity. This is an energy caused by "moving charges" vital to technology. This energy comes in kinds called positive and negative charge.

Magnetism. This is an energy related to electricity that has two poles of North and South. The Earth is known to have magnetic poles.

Gravity. This is an energy whereby mass objects fall down.

Strong Force. This is a force that holds the nucleus together.

Motion. This is the action of masses changing location in space.

Rotation. This refers to masses that spin around.

Fire. This is the action of burning called combustion.

Heat. This is a phenomena involving sensations of warm and cool. It is studied in thermodynamics.

Friction. This results when surfaces collide and rub together.

Radiation. This refers to all manner of energies like radio, light, infrared, ultraviolet, X-rays, and gamma rays.

Tides. These are dramas where the Moon's gravity pulls water in. The Bay of Fundy in Canada is an epic place to learn about tides.

Sound. This is an energy born of bangs and shocks to hear by.

Vibration. This refers to "sound waves" traveling in "mediums" like rock.

Effects. Nature is filled with many kinds of disturbances involving energy and mass and are thus known as 'effects'.

The Universe. This is the totality of all things. This is the sum total of space, time, mass, and energy known to physics.

Stars. These are mass objects known to glow in the sky. They come in many kinds ranging from giants to dwarves. The Sun is known to be a star.

The Sun. This is the home star of the solar system and the most studied star in science. It is studied in its own science called heliophysics.

The Moon. This is the natural satellite of the Earth.

The Earth. This is the home planet for Man, life, continents, and much more.

Planets. These are mass objects known to gravitationally "orbit" the Sun. Presently eight are known to exist.

THE GRAND SCIENCE OF PHYSICS

Fluids. This refers to substances that flow or are liquids.
Plasma. This refers to gases heated where they are ionized and glowing.
Ions. This refers to atoms and particles that gain a charge.
Waves. These are dramas of "undulation" concerned with ideas like frequency. Ocean waves, ripples, EM waves, and so on are all examples of waves.
Particles. These are very small mass objects like electrons, protons, and neutrons.
Atoms. This refers to very small particle systems consisting of a nucleus and electrons.
Elements. This refers to the many kinds of atomic substance each differing by the mass of its nucleus.
Black Holes. These are stars crushed by gravity to become "bottomless pits that crush." Their gravity is so strong that light cannot escape them.
Cosmic Rays. These are energetic kinds of radiation known from space.
X Rays. This refers to an energetic kind of "light" discovered by W. Roentgen.
Radio. This refers to a kind of "light" known for radio broadcasts.
Planetary Motion. This refers to the movement of planets about the Sun.
Aurora. This refers to "electrical disturbances" in the atmosphere (called Northern Lights), also called aurora borealis.
Solar Wind. This refers to light, heat, particles, and energies emitted by the Sun.
Electromagnetism. This is a union of electricity with magnetism.
Radioactivity. This is the "energetic self-destruction" of atomic nuclei.
Weather. This refers to phenomena consisting of air, wind, energy, and other qualities.
Quakes. This refers to shaking of the mass of Earth due to tectonic plate movements.
Flight. This refers to how airplanes and birds act to fly in the sky.

There are in reality many kinds of phenomenon, many are obscure subjects in physics.

PHYSICS

"A force is a power to attract, repel, cause motion, and show fields exist."

-class discussion

Existence

This is the idea of what is actual, real, physical, of Nature, and can be sensed by Man and living things. It is the task of physics to identify what is real in Nature and study it. Physicists have found that space, time, mass, energy, X rays, inertia, weight, cosmic rays, stars, planets, and so on exist (they are acknowledged to be real in physics). Physicists have realized that time travel, perpetual motion, orgone, pyramid power, odic force, N rays, and mana do not exist (they are acknowledged not to be real in physics). Existence is an intense issue to discuss and is a main topic in science, physics, and philosophy.

Experience

This word refers to what can be sensed and known. An experience can be watching clouds, feeling wind, touching fur, walking down the street, and so on. If the things of Nature could not be experienced, they could not be studied in physics.

Force

For ages, various thinkers thought Nature was filled with many kinds of "forces" or powers of energy to cause movement. Examples were light, heat, electricity, magnetism, friction, centrifugal force (spinning things), gravity, Coriolis force, cosmic gravitational force, material force, nuclear force, atomic force, and so on. There were a vast and confusing number of them and scientists did not know what caused them. Eventually, various thinkers investigated them and found that many of these forces were byproducts of still more fundamental forces. Today, nature's diversity has been reduced to the actions of just five (or three) fundamental forces.

Fundamental Force

A fundamental force is a force whereby many other forces are said to originate from or be reduced to it. An idea of this is ice, snow, sleet, hail, water vapor, steam, and so on. They can be thought of as "forces". It is well-known that all these things are as aspects of water. Water then can be said to be a "fundamental force". Presently, physics knows four of them in gravity, electromagnetism, the weak nuclear force, and the strong nuclear force. All of today's natural forces have been reduced to actions of just these four forces. Today, its known that electromagnetism and the weak nuclear force are the results of the action of still a more basic force in the electroweak force. Presently, it can be said there are just three fundamental forces in gravity, the strong nuclear force, and the electroweak force. There is a quest in physics to reduce all these forces to aspects of a single, fundamental force.

Field

Fields are energy presences in space created by mass. Types of field are the electric, magnetic, electromagnetic, and gravitational fields. Fields are created by mass and mass qualities called "mass", charge, and so on. Fields are created around all mass objects like stars, planets, electrons, protons, nuclei, and so on. When particles are within the presence of a field, movement may be caused by a force. Fields are able to attract, repel, and cause motion. Fields cause forces to exist.

Fundamental Forces

Physicists (or the scientists of physics) have found that Nature has five "basic" forces or effects. These forces are energy influences by which material things "feel" or act accordingly to in some way. Their discussion tends to be technical and fill many textbooks. Examples are the following:

Gravity: This is a force responsible for causing mass objects to fall and keep planets in their orbits. It is currently described by a theory called general relativity and for centuries has been an enigma in

PHYSICS

science. It was thought there were two kinds of gravity, a celestial (cosmic) and terrestrial (of Earth). Isaac Newton came along and thought there was only one, true kind of gravity and its known today as "universal gravitation". Today, gravity still hides many secrets and physicists do not understand it completely.

Electromagnetism (EM). This is a force that is a union of electricity and magnetism as one force. It is responsible for such energies like light, heat, radio, X rays, gamma rays, and so on. A theory by the famed physicist, James Clerk Maxwell united these forces and their rules are called Maxwell's Laws.

Weak Nuclear Force (WNF). This is a force responsible for decay in atoms.

Strong Nuclear Force (SNF). This is a force responsible for binding nuclei together.

Electroweak Force (EW). This is a force that is a union of electromagnetism and the weak nuclear force. Discovered by the trio of Weinberg, Salam, and Glashow, it is the newest fundamental force known. It can be said there are now just three fundamental forces in Nature or just the forces of the electroweak, gravity, and strong nuclear force.

Reductionism

This is the activity of trying to reduce things to a "simplicity" or to try to make simpler a vast amount of complexity. An example of a complexity is the wide array of "things" called hail, ice, sleet, rain, water vapor, fog, clouds, steam, and so on. Its known in physics that all these things are just "forms" of water and they have been reduced or made simpler of understanding by saying they are forms of water. Physicists are on the search for "simplicity" in Nature by trying to reduce Nature's complexity to simpler things like fundamental forces.

Education in Physics

Physics is a very difficult subject to study. It takes years of effort and is for students who like to read books, attend seminars, and associate

THE GRAND SCIENCE OF PHYSICS

with those with science educations. A brief discussion on what its like to "major" in physics goes like this:

- An education in physics usually begins in high school taking courses in calculus, analytic geometry, trigonometry, physical science, chemistry, physics, etc.
- Students should get very good grades (B to A), maintain high GPAs, like to study science, and just have a "passion" for science.
- Students should go to colleges that offer physics "majors" or a dedication to coursework in physics.
- Students in their first college year will take calculus and chemistry. These subjects go hand in hand with physics as they intertwine immensely.
- In the next year, students will take formal physics courses in electromagnetism, mechanics, "pre-engineering", optics, and other subjects.
- In their third year, they may take "higher math" and quantum physics, relativity, electronics, and other "higher physics" courses.
- In the fourth year, they will complete advanced physics courses and complete all credit requirements to get a "bachelor of science" degree. This certificate says they have completed a physics major in college.

Physics Professor

Some students of physics will go on to graduate school, obtain a master's or doctorate degree, and attain to the rank of "professor". A professor will usually have a teaching position in a college, have specialized in a branch of physics, conducts physics research, and in some way "professes" expertise in physics. Physics professors are physics' main academic authorities, explorers of the unknown, trainers of future physicists, and main embodiments of physics expertise.

PHYSICS

Physics Institute

Physics is taught worldwide and is found almost everywhere. It is a fascinating science exploring Nature for phenomena, natural laws, technology, effects, and more. Physics has many sub-disciplines and thus many kinds of physics institute has been so founded to research such branches. Physics is practiced at institutes like the following:

Observatories
Particle Accelerators
Nuclear Reactors
Neutrino telescopes
Volcano observatories
Radio telescopes
Electronics laboratories
Weather stations
Weapons laboratories
Cosmology institutes
Space agencies
Bomb Test Sites
Physics Colleges

Laws of Physics

Physics is filled with numerous rules expressed as equations. These rules are vital to any education in physics. They are statements by which physical phenomena are said to obey. In the ordinary world, everyone is familiar with laws like obey the speed limit and do not steal, the same is also true of Nature. Here phenomena such as light can only travel at one speed or act according o precise rules. As to why phenomena obey laws no one knows, however it is known that there are many such laws and an education in them is vital to physics.

Physics equations are complex statements about how Nature behaves. They are expressed in the language of mathematics and require a good education to understand and use. As a service to readers with no math background, here is a basic introduction to what physics equations are.

THE GRAND SCIENCE OF PHYSICS

There is in mathematics ten basic numbers such as the following:
0 1 2 3 4 5 6 7 8 9

Each number represents an amount. Examples are one is only a bit. Two would represent a double of a bit, zero represents nothing of anything, and so on.

There are various operations by which numbers are related to each other. Examples are multiplication, division, subtraction, and addition. They are expressed in the symbols *, /, -. +.

Numbers can be related to each other by the following:

One added to one is two or expressed mathematically 1+1 = 2

Five multiplied by three is fifteen or expressed mathematically as 5*3 = 15

Nine divided by three is three or expressed mathematically 9/3 = 3

Six subtracted from seven is one or expressed mathematically 7-6 = 1.

Each mathematical expression is called an equation or statement of equals to.

Equations are used throughout mathematics and can go on to become very complicated things.

In physics, there exists such ideas as mass, energy, space, time, force, power, work, and so on.

These ideas can be expressed as numbers in the form of letters.

Examples are energy is expressed as the number E. That is E can amount to a number like 0, 1, 2, 3 and so on. Physicists however do not know the value for the amount of energy and hence just list it as E. Physicists have taken to devising equations where numbers have been replaced with letters. Physics is filled with such equations that have gone on to be famous in their own right. Examples are

E = m * c * c, F = m * a, F = G * M * m/r * r, W = m * g, and E = h * f.

The most famous equation physics knows was proposed by the famous Albert Einstein. He stated that energy is mass multiplied by the value for the speed light multiplied yet again by the value of the speed of light or the famous expression E = mc (squared). In physics, there are many such equations and some are very difficult to discuss. Whole textbooks have been published on physics

PHYSICS

equations and their implications and their study is vital to the science of physics.

Physical Laws

Physics has many kinds of law or rule that describe how phenomena behave. Just as there are rules to say do not steal, lie, murder, and cheat, Nature has rules by which it follows. Examples of laws known to physics are:

Newton's law of gravity. This is an equation that says masses "exert" a gravitational force on each other, A mass has a "field" of energy about that "pulls" other masses closer to it. This equation uses a number called G that allows one to calculate the force of gravity. An expression is $F=GMm/r^2$.

$E=mc^2$. This is perhaps the most famous law of physics. It says that mass is a kind of energy and within mass there are vast amounts of energy.

Hooke's law. This is a law about springs (or coils of wire) and the force they have.

Ohm's law. This is a law named for George Ohm about voltage is a function of resistance and current. It is central to studies of electricity and is taught to all physics students in college. Its expression is $v=ir$.

Hubble's law. This is a law that calculates the speed of galaxies as the universe expands. Its expression is $v=Hr$.

$V=dx/dt$. This is a law that says velocity or speed is related to one's position in space. It is learned in studies of motion and is a basic rule of physics.

Momentum law. Momentum is the idea that moving masses like a car have an extra push to keep moving until they come to a stop, there is a famous equation calculating momentum in physics. Its expression is $p=mv$.

Laws of motion. These are rules by which mass objects are known to obey. Isaac Newton composed three of them and they are technical discussions. The reader is encouraged to explore these powerful statements taught throughout physics.

Maxwell's laws. These are laws by which electromagnetic waves obey. They were created (or "formulated") by James Clerk Maxwell in the 19th century

Combined gas law. This is the equation denoted by PV=nRT. It discusses gas behavior and is taught in physics and chemistry.

E=hf. This is a law stating that energy is related to frequency (Max Planck).

Laws of planetary motion. Johannes Kepler composed three laws by which planets are known to move about the Sun. They are complicated discussions so vital to space research today.

Laws of heat. Various thinkers composed laws that govern the behavior of heat. There are four of them and can be studied in the subject called thermodynamics (the changes of heat).

Units

There is in physics ideas called units. They are used in calculations and are vital in physics. Examples of units are the following:

For the idea of mass, there is the gram or a basic amount of mass.

For energy, there is the "joule" (jool) or a basic amount of energy. Other kinds of energy unit are the electron volt, erg, kilowatt hour, and so on.

For time, there is the second or basic amount of time. Other "amounts" are hours, days, weeks, years, decades, centuries, leap year, millennia, eons, eras, epochs, etc.

For space, there is the meter or foot, basic amounts or measurements of space.

Each unit can be subdivided or increased by many factors to make for even larger units. Units can be very complex things and their discussion tends to be highly mathematical.

Measurement

Physicists are always trying to find values and amounts. They want to know how much mass is in stars or how energetic a campfire can be. In so doing, they have devised ways to measure such things using tools and calculators. They try to "measure" or determine the amount of some number be it mass, energy, amount of space, amount of time, intensity of force, and so on. Measurement refers to using such things as rulers, metersticks, clocks, or other such device to find a value for numbers like energy, time, and so on. It is overall a vital practice in physics.

Physics Technology

Physics uses many kinds of device to engage in the practice of physics. Many were invented ages ago and some quite recently. They are all today used as part of the "arsenal" of physics tools used by professional scientists. Some key inventions used widely in physics are the following:

- Telescope. This is a tube with two lenses on either side. This allows for seeing things far away in a better way. Invented by such people like Lippershey, Galileo, and Newton, it is now an indispensable tool found in observatories around the world. Famous telescope installations are called the Keck, Subaru, Mt. Wilson, Yerkes, and others. A giant version has been placed in orbit and is called the Hubble Space Telescope.
- Radio telescope. This is a device shaped like a gigantic dish. It is a kind of telescope using radio waves invented by Reber and Jansky. Today, its found at many installations called Arecibo, Green Bank, the Very Large Array, and so on.
- Optical equipment. This refers to such tools of light as the lens, mirror, diffraction grating, prism, laser, gravitational lens, kaleidoscope, and so on.
- Atomic/nuclear technology. This refers to such devices like nuclear reactors, atomic bombs, atomic lasers, Geiger counter, atomic battery, and so on.
- Spectroscope. This is a device that measures colors to identify mass.
- Scale. This is a tool to measure weight and mass.

- Pendulum. This refers to a weight attached to a string and it swings back and forth. It is found in clocks and contraptions and a major thinker in them is named Leon Foucault.
- Spring. This is a coiled "line" of metal. It is studied in an issue called Hooke's Law.
- Laser. This refers to a beam of light able to burn holes in materials. Invented by such people as Townes, Basov, Prochorov, and others, it is a modern marvel found in technology like DVDs, CD players, laser eye surgery, and so on.
- Particle accelerator. This device originally called the cyclotron was invented by the Nobel prize winner Ernest Lawrence. It has since been adapted to build gigantic machines like the Large Hadron Collider, linear accelerators, and so on. Famous installations using them are called CERN, Fermilab, SLAC, DESY, and others.
- Crookes' tube. This is a famous device that allows for glows. It is a classic invention that lead to devices like light bulbs, neon lamps, CRT, TV, monitors, and so on.
- Leyden Jar. This is a famous device that leads to the discovery of electricity, batteries, and other electrical devices and is known from the Renaissance.
- Ionization chamber. This is a device that causes atoms to ionize (lose, gain electrons).
- Speedometer. This refers to devices that measure for speed and motion.
- Cathode ray tube. This device is the precursor of the television set and varieties of it are called the oscilloscope, monitor, and so on.
- Radio. This is a device using radio energy. It was pioneered by men like Marconi, Hertz, and others. Today, its found in radio stations, radar, and in other inventions.
- Charge couple device. Invented by the duo of Smith and Boyle, Nobel prize winners, it is found in photography and many sensor devices.

PHYSICS

- Electronics technology. This refers to devices going by the names of LED, battery, capacitor, transistor, transformer, diode, and so on.
- Pressure chamber. This device is known for causing crushing pressures and is used in high pressure physics.
- Wind tunnel. This is a tunnel with a giant fan to produce exceptional gusts of wind.
- Airplane. This device invented by the Wright Bros uses physics and engineering. Invented about 1900, it is today found worldwide and in many versions.
- Computer. Invented by such people as Babbage, Atanasoff, Berry, Shannon, Turing, and many others, it is a device used for calculating in large numbers. It is now an indispensable machine used throughout physics.
- Centrifuge. This is a device that spins rapidly illustrating centrifugal force.
- Metal detector. This is a device able to detect for buried objects.
- Celestial sphere. This is a "construct" with the Earth at the center and the "heavens" arranged about it according to latitude and longitude.
- Gravitational lens. This is a "tool" in astronomy whereby stars bend light to allow for photographs of distant astronomical objects.
- Freezer, refrigerator. Offshoot inventions of heat, they are used to cool masses to freezing temperatures.
- Heat technology. This refers to all kinds of invention using heat like ovens, thermocouples, irons, stoves, warmers, freezers, and so on.
- Vacuum chamber. This is a device that tries to simulate the vacuum of space. A vacuum cleaner is an offshoot of this device.
- Acoustic equipment. This refers to devices that cause sound like beats, music, amplifiers, radio, and so on.

- Simple machines. This refers to such inventions as the pulley, balance, screw, lever, wheel and axle, inclined plane, wedge, and so on.
- Detectors. This refers to machines able to sense or detect for energies and particles. Versions of them are the Geiger counter, cloud chamber, and bubble chamber.
- Space probe. This refers to satellites placed in space to observe the Sun, energies, particles, stars, planets, weather, and so on.
- Laboratories. These are places where physics research is conducted.
- Periodic Table of the Elements. This is a chart developed by the chemist Dmitri Mendeleev listing all the known elements. It is vital for use in chemistry, but also in physics. Even now, new elements are being discovered and it is being improved.
- Clock. This refers all kinds of device that measure time.
- Thermometer. This refers to a device that measures heat.
- Weather technology. This refers to devices to study weather like the barometer, bolometer, wind speed gage, and so on.
- Electromagnet. This is a mass of metal where an electric flow causes a magnet to form. It is used in recycling plants, electronic devices, and elsewhere.

There are in reality many kinds of invention, some famous while others quite obscure used in physics. Please read about others in physics textbooks.

Event

Events are the ideas of "occurrances" or "happenings" in Nature. Examples are the bounce of a ball, the smash of a glass, the hit of a fist, the firing of a gun, the blast of a bomb, an eye blink, a finger snap, a flash of light, a vibration, a beat of a drum, and so on. Events are deeply studied in physics as they factor in the discussion of motion.

PHYSICS

Action, Reaction

In physics, there are two classic ideas that are versions of the word event. An action can mean anything from the bounce of a ball to a snap of a finger. A reaction is an event that "responds" to what an action is.

Action At A Distance

This is a controversy associated with Newton. It is the discussion of how does a force "know" that an action is happening across space. It is an ancient issue and one that continues today in a subject called nonlocality.

Change

This word means to become different with time. It is a vital issue in physics found in discussions on motion, cosmic acceleration, process, and the flow of time. It seems like everything in Nature changes and change seems to be the only constant. Please read more about this word in other books.

Motion

In physics, scientists have taken to studying how masses move or change position in space. Called motion, it is an ancient discussion on the nature of moving things. Physicists know of several sets of rules or laws of motion. Examples are Kepler's laws of planetary motion, Newton's laws of motion, the laws of black hole motion, and so on. Motion is an idea that discusses the related notions of velocity, acceleration, speed, direction, rate, acceleration due to gravity, and so on. The reader is encouraged to explore more explanatory books on motion.

Concepts of Motion

Motion is a fundamental issue in physics filling countless books. Its study is vital to many sciences and a deep study of its basic words is required to understand motion. Various words to explore are the following:

- position –displacement –velocity –speed –acceleration –jerk –centrifugal force –centripetal force –wave motion –angular motion –spin –the speed of light –Coriolis force –torque – momentum –impulse –inertia –laws of motion –force –power –work –intensity –escape velocity –acceleration due to gravity –frame of reference –vector motion –frequency –period –wavelength –amplitude –angular velocity –moment of inertia –center of mass –center of gravity –specific gravity –electromotive force –friction –Doppler effect –collision –normal force

MOTION

–vector motion –nonmotion –pressure -temperature –air resistance –free fall –slip and slide –average velocity –pulse -rigidity -length contraction -relativistic motion -quantum jump -speed of gravity -coherence -phase -pitch -volume

Please read about the the nature of these words in physics textbooks.

Basic Motion

Motion is a complicated discussion in physics. It refers to mass objects changing position in space with passing of time. Discussion on its basic ideas are:

- Position. This is the idea of a "location" or "place in space".
- Displacement. This refers to a mass object moving in a certain amount of space.
- Velocity. This is the idea of a mass's rate of change of its location in a direction. A velocity can be the speed of light, five miles per hour, 55 miles per hour, and so on. Speed is another name for velocity although without reference to direction.
- Acceleration. This is the idea of a mass's velocity changing with time.
- Force is the idea of a "push or pull" causing motion.
- Non-motion. This is the idea of no motion, lack of movement, and so on.

Mechanics (or Kinematics)

Motion is studied in this subject which is taught in college. It is sub-divided into various branches of interest like:

- Statics. The study of objects not in motion.
- Dynamics. The study of objects in motion.
- Celestial mechanics. The study of the motion of astronomical objects.
- Orbital mechanics. The study of mass objects moving in orbits.
- Electrostatics and Electrodynamics. The study of the motion and non-motion in electric fields.

- Magnetostatics and Magnetodynamics. The study of motion and non-motion in magnetic fields.
- Aerodynamics. This is the study of the motion of air and gases.

Non-Motion

This is a word that refers to masses that do not move. It can also mean masses where the forces acting on them are all cancelled out or appear not to exist. Non-motion is studied in statics, a branch of mechanics dealing with non-moving masses.

Laws of Motion

Motion is known to follow rules called the "laws of motion". There exists many kinds of laws of motion, however there are three that are considered the most famous. They were conceived by Isaac Newton and are standard teaching throughout physics. A discussion on them goes as follows.

- First Law. This refers to mass objects that once set in motion moving in a straight line will continue to do so essentially forever, however mass objects can be changed of their paths if a force is applied to them.
- Second Law. This refers to the statement, $F=ma$. It says that a force is the result of a mass with an "acceleration" applied to it.
- Third Law. This refers to the classic statement "for every action there is an equal and opposite reaction". What this says is that for actions like pushes and pulls, an equal force in the opposite direction will occur. A tug of war illustrates what this law is about.

The laws of motion are discussed in many physics textbooks. They are among some of the most famous expressions known in science. Please read about motion in physics textbooks.

Zeno's Paradoxes

In Ancient Greece, there lived a thinker named Zeno of Elea. He is credited with conceiving "thought experiments" discussing motion.

MOTION

Various paradoxes discuss the turtle and the hare in a foot race, the action of completing a journey, but going half way, and other discussions. Zeno's paradoxes (as they are called) attempt to argue that motion does not exist, that time is not real, that a journey cannot happen, and so on. They have fascinated thinkers for ages and are part of the study of motion.

Angular Motion

This word refers to the discussion of "spinning objects" or objects that move about an "axis" or line. Objects demonstrating angular motion are merry go-rounds, the Earth, planets, galaxies, stars, pulsars, gyroscopes, clocks, tops, centrifuges, yo-yos, and so on. Most anything round and that can spin exhibits this type of motion. Concepts associated with it for study are:
- angular speed –angular momentum –angular acceleration –centrifugal force –centripetal force –angular velocity –elliptical motion –planetary motion –orbits

Please read about the nature of these words in physics textbooks.

Wave Motion

Waves refer to "undulations" in a substance like water or some other fluid. Ripples, ocean waves, and so on all illustrate what waves are. Waves have been intensely studied in physics and are today a basic teaching. Its thought energies like light and heat travel in waves and subatomic particles have a wave nature. Various concepts of waves to study are the following:
- amplitude –wave speed or wave velocity –phase –frequency –wavelength –period –angle –crest –trough –wave energy –wave/particle duality (wavicle) –interference –diffraction –gravity wave –superposition –partition of waves –sine –cosine –tangent –oscillation –damping –fundamental frequency –secant –cosecant -cotangent

Please read about the fascinating issue of waves.

Projectile Motion

This refers to discussions on the motion of projectiles. A projectile can be anything from a cannonball launched from a cannon to a bullet fired from a gun. Projectiles fall in a path called a "parabola". Projectiles have been studied for ages and are a main topic in physics classes.

Brownian Motion

Robert Brown, a 19th century thinker observed grains floating in water and moving. From this observation, the issue called Brownian motion was invented in physics. Later on, Einstein would examine Brownian motion and publish a paper that Brownian motion suggested that atoms existed.

Situations of Motion

Motion is heavily studied in physics. It is a classic subject occupying the attention of thinkers through ages. Presented here are discussions on "regions of existence" where motion takes on intense discussion and interesting issues.

- Absolute Zero. This is a name for the coldest possible temperature. Here liquids and substances exhibit behavior unlike any found in Nature. Here masses become superconducting (allow electricity to flow without resistance), become superfluids (become fluids that can flow uphill), and assume structures like the Bose-Einstein Condensate. This "region" is currently studied in physics with many epic discoveries being found.
- High pressure. This is where masses are subjected to pressures of many tons.
- High temperature. This is where masses are heated to temperatures of hundreds to millions of degrees. Here masses move very fast and conditions like this are found in plasmas (ionized gases), inside stars, and inside planets.
- Near the speed of light. This is the study of masses in motion near the speed of light. Here space contracts and time dilates, energy of motion is converted into mass.

MOTION

- At the speed of light. It is known no mass object can travel at this speed, but energies are known to do so however.
- Faster than the speed of light (called superluminal or FTL). As of now, it is not believed anything travels faster than the speed of light. But speculation on such issues like tachyons, nonlocality, and inflation make this a lively area of discussion.
- Inside neutron stars. Inside these interesting kinds of star, matter is crushed under enormous pressure where a teaspoon of matter can weigh as much as the planet Earth.
- Inside black holes. The inside of a black hole is a discussion of fantasy and speculation. Here mass is crushed into a point called a singularity, a state of infinite pressure and density. It is unknown what matter is like here, but knowing it is sure to be interesting.
- Near a black hole. Its known a black hole's gravity is so strong that light cannot escape. Light will form a ring around a black hole and become trapped by the hole's gravity.
- Inside atoms and nuclei. Here inside these objects, matter behaves according to "quantum rules" and the ordinary world seems irrelevant.
- Inside stars. Here matter is heated to millions of degrees and behaves in highly energetic ways.
- In orbit. This refers to mass objects that go about a parent mass object. Examples are the planets going about the Sun, the Moon going about the Earth, a "moon galaxy" going about a "parent galaxy, and so on. Motion here is studied in a subject called orbital mechanics.
- Tides. Tides are "flows of water" caused by the Moon's gravity. Tides refer to water coming on to land and then going back to the sea by motion of the Moon. The Bay of Fundy in Canada is a famous place to study tides and their motion.
- Mass tide. The Moon's gravity pulls on the Earth and this produces the phenomena of tides. Types of tide are water (Bay of Fundy), air (gas streams), and land.

- In harmony. There is in physics an issue called "simple harmonic motion". It refers to things moving in harmony (or synchronicity) with each other. It is regarded as one of the easiest kinds of motion to learn about.
- The universe. Its known that the universe began in an explosive event called the Big Bang. It was discovered to be expanding and is not a changeless thing. Today, its believed the universe is accelerating of its expansion. It demonstrates "motion" unlike any other object and for reasons unknown to physics.
- Friction and collision. This refers to when surfaces collide or rub together and objects hitting each other.
- Rectilinear motion. This refers to masses moving in a line.
- Curvilinear motion. This refers to masses moving in "paths" (trajectories) consisting at times of lines and curves. Comets moving about the Sun and planets in orbit about the Sun illustrate this type of motion.
- Rigid body motion refers to masses that move, but do not change shape as they move.
- Resonance. Resonance refers to waves inside an object moving back and forth in harmony. It is an intensely studied subject in its own right.
- Nutation. This refers to the bobbing up-and-down of spinning tops.
- Oscillation. This refers to swinging back and forth or waves doing the same in moving back and forth.
- Periodic motion. This refers to objects in motion that repeat their movements in time.
- Fluid and gas motion. This refers to how gases and fluids move or when gases and fluids are mixed together.
- Elastic. This refers to mass objects being "deformed" and then reassuming their original shape.
- Free fall. This refers to mass objects that are dropped from on high and its discussion. Kinds of free fall are asteroid impacts, skydiving, dropping things, and so on.

MOTION

- Faster than the speed of sound. This refers to mass objects that travel faster than sound (called "supersonic"). At such speeds, noise effects called "sonic booms" occur that can break glass and windows.
- Air flow. This refers to air in motion like wind, the jet stream, storms, and so on.
- Water flow. This refers to water in motion like tides, currents, ocean waves, tidal waves, eddies, river flow, water falls, upwellings, and so on.
- Circular motion. This refers to movement in a circular path (circle). Being on a merry go-round, a spinning top, or spinning a yo-yo illustrate this type of motion.
- "Psychokinesis/Telekinesis??!!" There is in physics a controversy over an issue called psychokinesis, telekinesis or simply PK. It refers to causing motion by mind or by use of paranormal power. It is not known to be real and it is considered to be in disrepute in physics. Its discussion is age-old and filled with controversy. Many scientists have sought to verify it exists, study it, or prove it was real in some way. For now, it is a suspect issue in the study of motion, an issue at once part of pseudoscience, but tantalizing physicists that it could be real.

Perpetual Motion

There is a controversy in physics about perpetual motion. It refers to machines that once set into motion will stay in motion essentially forever without power supply. There is a rule in physics that says this kind of motion cannot occur. However, inventors throughout the ages have come up with devices that supposedly can do perpetual motion. Even now, many thinkers claim to have made such things. Today, any and all such claims are dismissed as 'garbage' in physics as no perpetual motion machine has been shown to be real.

Please read about this age-old issue in physics in other books.

Sound and Vibration

There is in physics discussion on situations of motion called sound and vibration. Sound refers to disturbances that lead to effects called beat, noise, music, and so on. Vibration refers to sound waves in mediums like rock, concrete, brick, ground, and so on. Collectively they are studied in a branch of physics called "acoustics".

Noise

Noise refers to sounds judged objectionable, hated, or disliked. Examples are jackhammers, yelling, screaming, fighting, cat meow, dog bark, rude sounds, and other displays. Noise can be said to be the opposite of music.

Music ("Of the muse")

Music is the idea of "pleasant sounds" that convey soothing, moods, emotional appeal, and entertainment. Its study is an offshoot of the study of sound. It has been practiced for ages by people like Mozart, Beethoven, Chaikovsky, Elvis Presley, Michael Jackson, Buddy Holly, and a host of others. Various branches of it are disco, rock, jazz, hip hop, opera, mood music, blues, reggae, classical music, pop, and so on. Many instruments have been created to engage in music like guitars, harps, banjo, drums, flutes, woodwinds, string instruments, and so on. It is a vibrant area of study, enjoyment, and exploration taught in schools worldwide.

Offshoots of Acoustics

Acoustics has grown into an intense science all its own. It has inspired many subjects each with its practitioners. Presented here are some of its derivatives:

- Hypersonics: the study of sound wave technology for transducers.
- Ultrasonics: The production and use of sound waves greater than 500 MHz.

- Supersonics: The development of aircraft able to travel faster than sound.
- Infrasonics. The study of sounds that cannot be heard.
- Megaphonics: The study of a tool called the megaphone (or loud horn).

Please study acoustics in more definitive texts.

Earthquakes

Its known the planet Earth undergoes "geologic (seismic) disturbances" that result in "earthquakes". Quakes can be disastrous dramas that can devastate cities and destroy communities. They are caused by volcanoes, fault movements, and other disturbances. They generate "seismic waves" that cause the ground to shake. They are studied in acoustics under vibrations. Terrible quakes known to history are the Alaskan, Chile, and the much feared future quake nicknamed the "Big One".

Weather

The Earth's atmosphere is made of gases, clouds, and has currents called wind. Together, they combine to form phenomena called "weather". Weather includes such things as wind, rains, storms, snowfall, clouds, sunshiny days, haze, fog, vog (volcano fog), smog (smoke fog), and so on. Storms are destructive dramas like hurricanes, typhoons, blizzards, tornadoes, waterspouts, and so on. Collectively weather is studied in the science of meteorology, a branch of physics.

Atoms (Atomos or smallest bit)

For centuries, physicists have wanted to know what the smallest particle of matter was. Various thinkers have taken to inventing notions like atomos or the idea of the smallest particle, it literally means in the Greek language to "not cut" or "has no smallest part". Since then, physicists have explored the notion of atom and have realized there are small particles called electrons, protons, and neutrons. Today, physicists do not know if there is a smallest particle, but the search continues. Major thinkers into the atom are Democritus, Dalton, Boscovich, Rutherford, JJ Thomson, Bohr, Goeppart Mayer, Fermi, Hahn, Meitner, and many others.

Nucleus ("Of The Center")

It was found by such thinkers as Rutherford, Chadwick, and others that atoms have a "center". This center has been named the "nucleus". Within a nucleus are particles called protons and neutrons (called nucleons) held together by the strong nuclear force. The nucleus represents the place where an atom is most massive. An analogy is the solar system. The Sun is thought to be the "center" of the solar system and is in many ways like the nucleus. Here in the Sun (nucleus), most of the mass of the solar system (and also the atom) is found. Its known that the nucleus is "orbited" by particles called electrons (similar to planets). Together, the nucleus and its associated electrons are what atoms are.

ATOMS (ATOMOS OR SMALLEST BIT)

Atom and Nucleus Models (Nuclear-"of the nucleus")

When atoms and the nucleus were discovered, they were found to be so small as to be invisible to sight. Various physicists have wanted to know what atoms and the nucleus look like. They have taken to inventing pictures or models of how to view them. It was thought the atom was something like the solar system, a dish of pudding, or something else. It was thought the nucleus was a ball of mass like a collection of marbles bound together. Today, these visions have undergone "revision" and now its thought atoms are like a solar system with electrons having the "chance" of appearing somewhere in orbit about the nucleus.

Chemical Elements

There is in physics and chemistry a chart called the Periodic Table of the Elements. It is a chart showing each and every kind of "chemical element" discovered and even now more elements are being added as they are discovered. A chemical element is an "atom" defined by its "nucleus". The lightest element is called hydrogen. Within its nucleus is one proton. The next lightest element is called helium. Within its nucleus there are two protons. After that, the next lightest element lithium has three protons. Each element in the chart has its own unique amount of protons in its nucleus that define the nature of an element. Elements can be combined together to form compounds, substances that make up the world like glass, plastic, DNA, concrete, nylon, and so on. The study of the elements makes up the science of chemistry.

Please read about this famous chart in science as well as the subject of chemistry.

Elements

In ancient times, it was thought there were only a few basic substances in Nature. Kinds of substance were Earth, Air, Water, Fire, Ether, Quintessence, Spirit, and Wood. Later on, thinkers began to question these beliefs and investigate the nature of these elements.

They found that many could be sub-divided into other substances and hence these elements could not be fundamental. In time, thinkers assembled a "chart" later to be called the Periodic Table of the Elements listing all the new and known elements that came from questioning. In time, the list of elements would grow into the hundreds. Presented here is a basic discussion on the "elements".

- Hydrogen. This is the lightest known element and is a gas that makes up water.
- Noble gases. This refers to elements named helium, neon, argon, krypton, xenon, and radon. They do not react with other elements and are gases.
- Neon is used for lighting.
- Radon is the heaviest known noble gas and is feared for its ability to cause cancer as it collects in houses.
- Halogens. This refers to elements from fluorine down to astatine. They may form gases and are known to be dangerous to "breathe".
- Salts. This refers to elements from lithium to francium which are known to combine to form salts.
- Francium (France) is the most reactive element known.
- Carbon. This element is known to make up living substances, diamond, graphite, buckyballs, amino acids, proteins, and other substances. It is fundamental to life and its known all living things are made of it in some way.
- Nitrogen. This element is known to make up the atmosphere and acids.
- Oxygen. This element is vital for breathing, life, burning, and so on.
- Iron. This metal was known to the ancients and was used in magnetism, steel, structures, money, rusting, sculpture, and other domains.
- Silicon is an element vital for use in making computers and electronics. There is a speculation that silicon could be used to make living matter.

ATOMS (ATOMOS OR SMALLEST BIT)

- Aluminum is a metal for making cans, money, and airplanes.
- Phosphorus is used in making acids.
- Sulfur is known for its noxious smell, acids, burning, and many uses.
- Mercury was known to the ancients as "quicksilver" and is a heavy liquid.
- Lead (Plumbum) was used in structures, is known as a poison, and alchemists tried to change it into gold. It factors in the activity of plumbing.
- Gold (Aurum) is a metal prized as money, wealth, for use in coins, for use in art, its yellowish shine, and its many uses. It is mined worldwide and the pursuit of gold as money has caused many "gold rushes".
- Silver (Argentum) is a counterpart to gold sought after by miners and used in photography. It has like gold been used in coins, caused greed, and so on.
- Iridium is a precious metal known to be found in asteroids.
- Tungsten (or Wolfram) is a metal famous for use in building.
- Technetium is named for technology and is a novel metal.
- Uranium. This highly radioactive metal is used to make atomic bombs.
- Plutonium. This metal is used in atomic bombs and exposure to even a grain of it can be deadly as its highly radioactive.
- Copper is a famous metal used in coins, wires, construction, and so on.
- Boron is used to make many kinds of chemicals.
- Arsenic is known to be a highly toxic poison.
- Tin is used in construction, cans, and many metallic creations.
- Metals. This refers to a vast array of elements (most of the elements) that have a "shine", luster, react with other metals, and can be used in construction.
- Precious metals. This refers to metals that have historically been used as money. Examples are platinum, gold, silver, osmium, rhodium, and palladium.

- Actinides. This refers to a wide array of radioactive metals.
- Transuranics. This refers to "beyond or heavier than uranium" elements called by names like americium, nobelium, curium, meitnerium, seaborgium, and dubnium.
- Rare earths. This refers to elements going from lanthanum across to lutetium. Historically, they were rare and derived from other substances hence "rare earth".
- Beryllides. This refers to elements from beryllium down to radium.

Please study chemistry to learn more about the nature of the chemical elements.

Molecules

Atoms can combine to form natural substances. Various kinds are acids, bases, water, rocks, minerals, gems (gemology), wood, plant matter, DNA, nucleic acids, amino acids, proteins, enzymes, catalysts, glass, metals, metalloids, and so on. There are many kinds of substance and each is studied in the science of chemistry.

Chemistry

Chemistry is an ancient science that had its origins in alchemy, the study of rocks, gems, soil, acids, water, bases, gases, liquids, and other substances. It grew over time where its scientists (called chemists) discovered many kinds of "element" (or fundamental substance). A leading thinker named Dmitri Mendeleev composed a chart listing all known elements (called the Periodic Table of the Elements). This chart forms the basic teaching of chemistry and its discussions. Chemistry in time would grow into a deeply complicated subject discussing acids, bases, pH, the "mole", corrosion, molecules, chemicals, gases, liquids, solids, and other issues. It has now grown so "labyrinthine", it has evolved into its own science, major, and profession. An education in chemistry is vital and highly useful in physics.

ATOMS (ATOMOS OR SMALLEST BIT)

Physics Theories

There is in physics various bodies of beliefs that were invented to "explain" or describe the behavior of various physical phenomena. Called theories, they are powerful constructions that allow physicists to investigate, think about, and explain natural occurrences. Theories in physics have the following qualities: a basic principle, a set of equations, the ability to describe, and the ability to predict. Theories have been invented to describe such things as heat behavior, light behavior, the behavior of atoms, the expansion of the universe, the behavior of gases, and so on. Famous theories known to physics are the examples: BCS theory, special relativity, general relativity, quantum mechanics, thermodynamics, classical mechanics, and so on. Their discussion tends to be technical and the reader is encouraged to explore them further.

Branches of Physics

Physics today is a vast labyrinth of disciplines that all share physics knowledge in some way. That is all these subjects require a physics education in order to understand and make a career in. Many of these subjects are sciences in their own right and are intense discussions filling many books themselves. Examples of offshoot subjects of physics are the following: astronomy, cosmology, astrophysics, solid state physics, electronics, condensed matter physics, quantum mechanics, relativity, gravity research, heliophysics, selenology, holography, engineering, particle physics, and so on. Various branches to explore are:

- Astronomy. This is the science of studying the stars and other cosmic phenomena. Originally joined with astrology, it broke apart and became its own science.
- Astrometry is the locating of the positions of stars in the sky.
- Planetology is the study of the planets.
- Heliophysics is the study of the Sun.
- Thermodynamics is the study of heat.
- Cosmology. This is the study of the universe as a whole, not the stars.

- Electronics. This is the study of devices of electricity like TVs, radio, computers, etc.
- Gravity research is the study of ways to cause antigravity.
- Gravity theory is the discussion on theories about gravity.
- Selenology is the study of the Moon. Selenography is the mapping of the Moon.
- Condensed matter physics studies issues like phonons, particles, and the like.
- Particle physics studies subatomic particles.
- Nuclear physics studies the nucleus of the atom, nuclear energy, and the like.
- Relativity is the study of Einstein's two theories of special and general relativity.
- Quantum mechanics studies quantum effects, nonlocality, the Bohr model, etc.
- Quantum reality is a discussion on what is reality really in quantum mechanics.
- Engineering is the study of constructions like skyscrapers, bridges, and so on.
- Architecture is the design of constructions built in engineering.
- Mechanics is the study of motion.
- Holography is the study of making three dimensional photographs.
- Photography is a technology into making pictures called 'photographs'.
- Optics is the study of light.
- Advanced physics is the study of such topics as superstring theory, M theory, etc.
- BCS theory is the study of the Bardeen-Cooper-Schrieffer theory of superconductivity.
- Physics history is the study of the history of physics people, events, and dramas.
- Acoustics is the study of sound.

ATOMS (ATOMOS OR SMALLEST BIT)

- Seismology is the study of earthquakes, their waves, shocks, and so on.
- Meteorology is the study of the weather.
- Metrology is the study of physics units.
- Geophysics is a blend of geology with physics.
- Biophysics is a blend of biology with physics.
- Astrophysics is a blend of astronomy with physics.
- Aerodynamics is the study of the motion of air.

There are in reality many branches of physics some famous while others quite obscure. Virtually all branches are "tough educations" and even now new branches are appearing. Physics is more than just studying motion, atoms, and light, it is really a collection of many sciences all collected under the word physics.

Traditions of Physics

Physics is a science filled with many practices and qualities. These practices grew up over centuries and have become something of a set of unwritten rules in the science. Examples are:

- Whenever a new law or constant is discovered, it is almost always named for the discoverer. Examples of this are Hooke's law, the Rydberg constant, Planck's constant, Newton's law of cooling, and the Einstein field equations.
- Whenever new planets in the solar system are discovered, they are always named for Roman gods. Dwarf planets are generally named for gods of other mythologies. Asteroids are generally named for anything. Comets are generally named for their discoverers.
- When new chemical elements are discovered, the finders generally have the right of name for it. Chemical elements are named by adding the suffix –ium after it. Examples are the elements seaborgium, meitnerium, nobelium, and so on. While not all chemical elements are named like this, it is a standard practice nowadays to name elements in this manner.

- Whenever new subatomic particles are discovered, they are named by adding the suffix –on to it. Examples are the proton, neutron, electron, photon, and so on. Sometimes if the subatomic particle is very small, then the suffix -ino is added, derived names are neutrino, axino, Higgsino, photino, etc.
- New effects of physics are generally named for their discoverers. Examples are the Aharonov-Bohm effect named for the duo of Aharonov and Bohm. Other kinds of effect are the Compton, Mossbauer, Cerenkov, Hall, Casimir, and Unruh.
- New physics discoveries are to be revealed to the world by being published in professional journals and being subject to peer review or analysis by professors.
- New branches of physics when they are discovered are usually named for what it is that is being studied. An example is "thermodynamics", literally meaning thermos and dynamics or the science of the changes of heat.
- New physics theories are generally named for their discoverer or what is being studied. Examples are Einstein's theory, the BCS theory, the Copernican principle, and so on.
- Awards are given to scientists who were vital to the discovery of some aspect of physics. The Nobel Prize is considered the "highest" award of physics and is given to scientists who made monumental discoveries in physics.

Organizations of Physics

Physics has grown so large of science, it now embraces such subjects like cosmology, astronomy, meteorology, and many others. Various organizations have grown up to represent or champion the cause of physics in some way. They may standardize practices, name units, publish journals, or in some way act as the "guardian" of physics or a sub-branch. Various organizations famous for doing so are:
- International Astronomical Union –American Institute of Physics –Royal Society –IUPAP –Astronomical Society of the Pacific –Academy of Sciences –American Meteorological

Association –Royal Astronomical Society –IEEE –Albert Einstein Society -AAAS

American Institute of Physics

In the USA an institute has arisen to champion the cause of physics. It is based in the state of Maryland and publishes many journals. It is a place where many physics organizations are based. It has a vast physics membership and acts as a kind of authority of physics in the USA.

Conservation Laws

There are in physics, certain kinds of laws that govern how energies and matter behave. They usually say in some way that the amount of some substance in the beginning is the same at the end. A famous kind of conservation law is the Law of Conservation of Mass. It briefly states that mass cannot be created or destroyed, but can only be conserved from interaction to interaction. Restated this means that the amount of mass you have at the beginning of a physical process is the same at the end. Conservation laws are vital to physics and their discussion tends to be intense.

Arrows of Physics

Physics is filled with many kinds of "arrow" or "directional processes that have been discussed. They are curiosities for discussion and some examples are:
- Heat arrow (from hot to cold) –Gravity's arrow (from up to down) –Time' arrow (from past to future) –Causality's arrow (from cause to effect) –Travel's arrow (from here to there) –Chemical arrow (reactants to products) –Electric arrow (from power to ground) –Black hole's arrow (from the universe across the event horizon to inside the hole)

Paradoxes

In physics, there are arguments that try to "trick the mind" or "discuss phenomena" in ways that seem confusing. Called paradoxes, they are age old discussions into the nature of physical reality. They require deep study, but are fundamental to learning physics. Examples to explore are:

- Zeno's paradoxes of motion –Olber's paradox –EPR paradox –Schrodinger's Cat

Experiments

There is in physics certain dramas called experiments. They are tests of truth as to the claim of a theory. An experiment is in some sense a trial to determine if something exists or not. For example, various thinkers have claimed particles like electrons and protons exist and experiments were devised to test to see if they actually existed. Famous experiments known to physics history are the following: the Michelson-Morley experiment, the gold foil experiment, the oil drop experiment, and so on. Their discussions tend to be intense and the reader is urged to explore them further.

Models

Physics studies "conceptions" or creations called "models". Models are "visions" of how something behaves, is represented, or is a "mirror of reality". An example of a model is a "globe" representing the planet Earth. A globe is a "representation" of what the Earth looks like, but is not the Earth itself. A map is also a kind of model as it represents geographic features. In all, models have their value as they are "pictures" of how reality seems to be. Physicists invent all kinds of models to study phenomena.

Principles

These are ideas in physics that refer to a claim or basic belief. Principles are vital to the understanding of physics theories. Examples of principles are causality, the anthropic principle, the equivalence principle, the Copernican principle, the holographic principle, the principle of relativity, and so on. Causality is a belief that causes cannot precede effects and another says that light can only go at one speed, nothing faster or slower. Physics is filled with many kinds of principles and they can be explored in other works.

Causality ("Cause and Effect")

This is an idea in physics that states causes precede effects. An example would be hitting a baseball. Before a bat can hit the baseball, the ball must be thrown. That is you cannot hit a homerun without first the ball being thrown. Causality is intimately linked to the flow of time and passage of events. The notion that physics must behave "causally" where causes (throwing a baseball, throwing a fist, pulling a trigger) precedes effects (hitting the ball, receiving a punch, or the happening of a gunshot) is referred to as "real world rightness". The notion that a homerun can occur before a baseball is thrown is thought to be impossible. This notion is involved with claims that time travel is impossible. That is the future occurs before the past or an effect can happen before its cause. Please explore deeper discussions on time travel.

Effects

Physics study is filled with many kinds of effects. Effects are the curious behavior of mass, energy, space, and time dramas. There exist many kinds called by the names of the Aharonov-Bohm effect, the Cerenkov effect (blue light glow), the Casimir effect, Auger effect, Doppler effect, Hall effect, and so on. Effects are studied and technology and inventions are made from them. A famous example is the photoelectric effect, it is the conversion of sunlight into electricity. The harnessing of this effect produces solar cell technology and was studied by Einstein. Effects tend to be extremely technical in their discussion and can be explored further in physics textbooks.

Greek Alphabet

The Greek alphabet consists of "letters" like alpha, beta, gamma, delta, all the way to omega. Physicists use these letters to name kinds of decay, subatomic particles, symbols for subatomic particles, and so on. Greek letters are found throughout physics and mathematics and are used for all kinds of purposes like defining functions or naming particles.

English Alphabet

This refers to letters going from ABCDEFG all the way to Z. Physicists use these letters to stand for physics "quantities". Examples are E for electric field, B for magnetic field, E for energy, m for mass, x for position in space, a for acceleration, v for velocity or speed, and so on. English letters along with Greek letters are used heavily in physics equations, particle symbols, and other types of uses.

Physics Variables

Physicists like to "quantify" concepts in physics (or make numbers of ideas). They tend to invent a symbol or use a letter to denote an idea and then use the "number" in an "equation" (or expression of math). Examples of physics variables are space (denoted x), time (denoted t), force (denoted f), acceleration (denoted a), velocity (denoted v), current (denoted i), voltage (denoted v), and so on. Physics variables are "unknown numbers". Physicists perform experiments and do measurements to find out what those numbers mean or what their values are.

Thinkers of Physics

There are in history certain thinkers who have greatly influenced the science of physics. They have gone on to propose theories, find constants, notice something profound, or contributed in some way to make physics what it is. Physics history is filled with their names and exploits and the reader is encouraged to explore their lives and careers. Examples of famous thinkers in physics to learn about are the following:
- Einstein –Newton –Feynman –Copernicus –Kepler –Brahe –Galileo –Faraday –Bohr –Aristotle –Archimedes –Pauli – Oppenheimer–Teller–Schwinger–Nambu-Hooke–Boscovich –Dalton –Heisenberg –Maxwell –Planck –Rutherford –Guth –Hawking -Curie –Roentgen –Young –Carnot –Clausius

ATOMS (ATOMOS OR SMALLEST BIT)

–Lorentz –Nobel laureates in physics –Wheeler –Hipparchus –Aristarchus –Alhacen –Raman –Weinberg –Townes –Dirac –Glashow

In physics, there are however some thinkers who have gone on to live lives bordering on fantasy for this science and the reader is especially encouraged to learn about Newton, Galileo, and Einstein. Many thinkers of physics have written books and papers that have gone on to have immense influence in shaping physics. Many famous examples are Newton's Principia, Einstein's 1905 annus mirabilis papers, Nobel prize-winning papers, and so on. The reader is urged to investigate these and similar works.

Unification

There is in physics five (or more basically three) known forces of Nature. However, there is an abiding belief that perhaps there is only one fundamental force of physics instead. As of now, there is an activity to try to unite the forces as one that is to make them all seem as if they were one force. Examples of famous unifications are Maxwell's theory where James Clerk Maxwell realized that electricity and magnetism were part of a unified force he called electromagnetism, Newton's law of gravity, the electroweak theory, and so on. There is in physics a modern attempt at trying to find one theory by which all the forces can be derived this is the search for the grand unified theory. An idea by which to better understand unification is the following example:

It is known that there are such things as ice, water vapor, hail, sleet, snow, and so on. They are in reality all just forms of water whether heated or frozen. In the nature of water, all these things are said to be "unified" or are aspects of the parent reality of water. In physics, it is thought gravity, electromagnetism, the nuclear forces, and the electroweak force are all just versions of a parent force. This notion as of today is only speculation in physics.

> "Physics today is in search of the one, the unity, the ultimate simplicity behind all the diversity of Nature."
>
> -class comment

Physics Revolutions

Physics knows the basic ideas of space, time, mass, and energy. Over the ages, there have been many "views" or "visions" of what these things are. Examples are that mass was a fundamental substance of Earth or that space was flat. Whenever the fundamental views of these words have changed, it is said a physics revolution has occurred. In 1905, Albert Einstein shook the world of physics when he said that space and time were not separate entities, but were a unity called spacetime and that mass was a form of energy. These proposals changed centuries old notions of what mass, energy, space, and time were thought to be. It is thought that perhaps our views of these ideas will change again leading to more physics revolutions in the future. A celebrated book by the author, Thomas Kuhn better explores the nature of revolutions and can be explored.

> "Thinkers of physics have believed radically different over time. If all the thinkers of the ages could be convened to meet and discuss physics with each other, there would be only insanity and confusion."
>
> -Voigt

Views of the Fundamental Concepts

Physics knows the basic ideas of space, time, mass, energy, spacetime, force, and so on. There have been many views of these things over the ages and even now, conceptions on them are still changing. Presented here are excerpts on how each concept has been looked upon.

Mass. It has been conceived as a divine substance, a magic, a fundamental substance of Earth, Air, Fire, and Water, a form of energy, an abstraction of mind, a primordial self-existent entity, and so on.

ATOMS (ATOMOS OR SMALLEST BIT)

Energy. It has been conceived as a divine emanation, a magic, a fundamental substance of Earth, Air, Fire, and Water, a spirit power, an abstraction of mind, a disturbance of ether, a primordial power, and the abiiity to do work.

Space. It has been conceived as a divine emanation, a magic, a fundamental substance of Earth, Air, Fire, and Water, an expanse of spirit, an abstraction of mind, a disturbance of ether, a creation of the void, a primordial entity, etc.

Time. It has been conceived as a divine power, a magic, a fundamental substance of Earth, Air, Fire, and Water, a consequence of space, a consequence of mass or energy, a disturbance of ether, a creation of the void, a primordial reality, a dimension of space or spacetime, an abstraction of mind, a non-existent entity somehow appearing as a reality, and so on.

Mass-Energy (also Massergy). This is the union of mass with energy as massergy, similar to space and time as spacetime. Massergy has been conceived of in ways similar to mass and energy as explained above.

Force. It has been conceived as a divine power, a magic, a fundamental substance of Earth, Air, Fire, and Water, a consequence of space or mass or energy or time, a primordial reality, an abstraction of mind, a disturbance of ether, an illusion of a deeper reality, and so on.

"Reality is what you think it is. If enough thinkers all believe that reality is just that, it is considered to be a theory, a finding, or a fact. Views of reality can change, facts can be understood in all sorts of ways. Its thought reality is a creation of the mind and that could very well be true."

-class comment

Physics Publications

Physics is a vast labyrinth of theories, facts, knowledge, and so on. The science today is too enormous for any one practitioner. However,

there are publications by which its knowledge is publicized. Examples are the following:
- textbooks –professional journals –popular science magazines –websites –books –newsletters –movies –TV shows –flyers – museum displays –captions –monuments –lecture notes –scrolls (ancient times) –newspapers –postings –press conferences

The reader is encouraged to read widely about the various developments in physics.

Physics Education

Physics today is one of the most complicated sciences around, it is taught in schools seemingly everywhere. Generally, professionals in physics start learning in grade school and advance through college to attain a degree in physics. From there, they study for a master's degree or doctorate and rise to the ranks of physics professors. Physics is a subject taught to kids early on and can take a lifetime to absorb and advance in. Physics is taught via lab demonstrations, homework, tutoring, and so on.

Physics Places

Physics is widely learned and explored in the following places:
- Colleges –institutes –observatories –particle accelerators –academia –national laboratories -technology companies –seminars –conventions –retreats –inventing companies –grade schools –high schools –planetariums –conferences –private industry – international associations –private laboratories

For a greater appreciation of physics, the reader is encouraged to visit some of these places where physics is advancing and physics is "happening" as to its new chapters of its history.

Physics Impossibility

There is in physics certain processes and events known to be impossible or are forbidden from occurring. Their discussion tends to be technical and stimulating and it can be explored further. Certain types

ATOMS (ATOMOS OR SMALLEST BIT)

of drama thought to be impossible in physics are the following:
- speeds faster than the speed of light –the flow of heat from cold to hot –time travel –perpetual motion –reversed causality –the spontaneous creation or destruction of mass and energy –the changing of a constant value –the violation of a physical law –the use of magic –paranormal effects –movement from inside a black hole to the outside -having a mass object travel at the speed of light -stopping time -creating time -destroying time -creating space -destroying space

Pseudophysics ("false physics")

This is a subject dealing with thinking about science fiction ideas as if they were "real physics". Ideas here are to make "real time machines", to use "energies" like mana or pyramid power for reactors, to build perpetual motion machines, and the like. It is a fun and speculative subject that blends science fiction, fantasy, and myth with physics to make for speculative ideas and discussions. However, this subject is not real physics, but is only an "imagining" of what physics could be or could be in fantasy.

"In ancient times, it was believed anything was possible in Nature. Magic spells, gods, world trees, miracles, divine powers, supernatural animals, love potions, magic places, forests of enchantment, magic swords, charms, and amulets are were thought to be real once. But as science advanced, these things were found to be impossible and physicists only want to believe in the real anyway."

-Voigt

Physics Subject Matter

Physics is an intense science filled with a vast world of intense ideas and developments. Examples are the following:
- Lasers –paradoxes –effects –theories –gravity –motion –heat –atoms –neutrinos –planets –stars -chemicals –superconductors

THE GRAND SCIENCE OF PHYSICS

–electricitiy – magnetism –spacetime –inventing –mysteries –light –black holes –the universe –spacetime –radio –sound – atomic energy –quarks –neutrinos –charge –quantum numbers –condensates –quantum mechanics –relativity –laws of motion –friction –work –weight –freefall –the Big Bang –the fate of the universe –orbits –tides –bosenova –nova –supernova –pulsars –hypernova -natural law -order -principle -paradigm -physical theory -galaxies -ether -spin -inertia -constants -variables

Please read about this and the myriad issues of physics.

Physics Textbook

Physics is a subject that has lasted centuries. There would exist many kinds of book that compile its teachings to be taught in school and these are called textbooks. Physics textbooks teach basic issues, laws, equations, principles, and more. They are technical things and require devotion and effort to understand. Physics overall is filled with literature discussing thinkers, effects, epic physicists, historical moments, and much more. Please visit a library to sample the many kinds of physics book.

Physics Department

In many colleges, there will be an academic department devoted to the science of physics. Here students can meet with professors, take classes, join clubs, and experience life in physics. Here professors will maintain offices, labs, libraries, centers, and more facilities maintaining the science of physics. Such departments often are lead by a leading professor known as a 'chair' and has a staff of assistants as well. Here physics finds a place where it is taught to students and the next generation of physicists find their training. Subjects like astronomy, cosmology, electronics, and other topics are usually taught here too.

Physics Club

In many schools there will exist a science club devoted to physics. Here students can join to discuss physics, to meet professors, to go on

ATOMS (ATOMOS OR SMALLEST BIT)

trips, to socialize, to make friends, and just to soak up physics. They are fun groups often overseen by a teacher as an advisor. More formal groups are the Society of Physics Students and Physics Honor Society.

Cultures of Physics

Physics or versions of it that discuss "physical reality" has been practiced for thousands of years. Many cultures had their own version of physics which was taught heavily influenced by cultural, religious, and other "biases". Physics as it is now known in the modern sense has only existed since the year 1900. Physics in spirit tries to understand the material world. Before physics appeared, ancient societies would try to understand the material world through creation myths, superstitions, fables, religious stories, and other accounts. While these stories would fail to explain Nature, in time they would have a legacy in the science of physics. Presented here are discussions on kinds of culture of physics or 'academic world' where people 'understood' Nature according to some belief system.

- Superstitious culture. Here "physics" is intertwined with beliefs in the occult, God, the gods, magic, mythologic cosmologies, omens, and spirits as an "understanding" of Nature. Here physics does not exist as a science, but superstition itself is thought of as academic teaching. In this time and ages like it, people believe in witchcraft, voodoo, sorcery, and all manner of superstitious belief. Even though later ages would discover such things are wrong, belief in them would be so pervasive that such people cannot believe any other way.
- Greek culture. Here physics is enmeshed with discussions on the gods, magic, ether, the four fundamental substances, the flow, logos, physis, cosmos, and other ideas. Greek philosophy is the primary resource as to teachings on Nature. A thinker named Democritus would invent the idea of the atomos later to become the notion of the atom. Pythagoreanism is a cult blending mysticism with geometry would fluorish. Aristotle would define this time in his writings. This is history's first

appearance of thinkers who would try to explain Nature without reference to magic or the gods.

- Ancient Egyptian culture. Here physics is intertwined with Egyptian gods, the idea of Ma'at (or divine order), the pharaoh, Nun, the cosmic sea, magic, and so on. The Egyptians would not know about physics, but they would know principles of engineering and beliefs similar to physics. The pharaoh or king would be seen as a god and absolute ruler over men.
- The 'Ancient World'. This refers generally to an era of history going from 5000 BC up to the fall of Rome near 500 AD. In this time, there would flourish societies like Ancient Egypt, Classical Greece, Babylon, Rome, Persia, Akkad, Assyria, Sumeria, Carthage, Phoenicia, Commagene, and other extinct societies. Science would not exist in this time. Instead beliefs about Nature would embrace ideas of magic, the gods, a supreme creator god, divine beings, and other myths. These beliefs would be replaced with science later on.
- Babylon. Here physics is intertwined with magic, the occult, the gods, and so on.
- Persia. Here physics is intertwined with the gods, magic, Zoroastrianism, and so on.
- Ancient China. Here physics is intertwined with ancient Chinese gods, the Tao, Yin and Yang, Confucian belief, chi (or energy), fundamental substances, and magic. Taoism and superstition are the main teachings in this age.
- Ancient Japan. Here physics is intertwined with ancient Japanese gods, magic, kami (spirits), fundamental substances, the worship of the Japanese emperor, and so on. Japan would borrow on ancient Chinese philosophy for its natural thinking.
- Ancient India. Here physics is intertwined with Jain, Buddhist, and Hindu beliefs. Atomism, Brahman, Vishnu and Shiva, the buddhanature, and so on are discussions. Ancient Indians would foreshadow many basic notions now found in modern physics.

ATOMS (ATOMOS OR SMALLEST BIT)

- Jainism. In India, there would grow up a religion similar to Buddhism. It would have its own worldview, ideas on atoms, and much more.
- Aborigines. Australian native tribes would maintain a worldview consisting of gods, monsters, and a mythical age known as Alchera or the Dreamtime.
- Eskimo. These Arctic native tribes would maintain myths on the gods and ice spirits.
- Maya and Aztec. Here physics is intertwined with the ancient gods of Mesoamerica, omens, spells, magic, the Aztec calendar, cycles of time, and so on.
- Inca. Here physics is intertwined with Inti, the Sun god, the Inca (divine king), magic, omens, seers, the god Viracocha, reverence for gold, and so on.
- Polynesia. Here physics is enmeshed with ancient Polynesian gods, mana (magic), spirits, omens, portents, taboo, huna (ancient religion), and superstition.
- Native American tribes. Here physics is enmeshed with prophecy, the occult, culture heroes, gods, magic, the vision quest, and so on.
- African tribes. Superstitions would dominate among African tribes.
- Rome. Here physics is like Greek culture as it is intertwined with magic, Roman gods, the occult, and so on. Lucretius is a major thinker here. Rome would discuss Greek philosophy.
- Vikings. In the Middle Ages, Norse barbarian tribes would believe the world was a tree (a world tree). Their view of Nature was dominated by superstition, Norse gods, magic, etc.
- Middle Ages. Here physics is intertwined with Bible teaching, Ptolemy beliefs, Aristotle's thinking, witchcraft, magic, superstition, the occult, and astrology. Natural philosophy was the academic teaching of this time. It is often described as a precursor to the science of physics to come. In this time astronomy and astrology would be seen as a union, a split would occur

where the science of astronomy would find its origin in this time.
- Arab Middle Ages. In the Middle Ages, Arab scholars would know of Greek philosophy. They would craft an academic world where Greek philosophy would blend with Islam.
- Renaissance. Here physics is assuming its modern form with Newton's ideas of absolute space and time, universal gravitation, laws of motion, ether, force, and so on. Physics would find its formal beginning in this time. Natural philosophy would become more refined and break with superstition. Epic thinkers like Newton, Galileo, Kepler, Liebniz, Descartes, Huygens, and others would contribute enormously defining the beginning of physics as a science.
- 18th century. In this time, the Renaissance would reach its climax in the Enlightenment. Thinkers like Benjamin Franklin and Boscovich would define this time. The Industrial Revolution would appear. Planet Uranus would be found by Herschel.
- 19th century. Here physics is like the Renaissance, but new ideas like disbelief in ether, Maxwell's laws, pre-relativity beliefs, and so on occur. Major thinkers of this era were Faraday, Oersted, Kirchoff, Maxwell, Volta, and many others. Planet Neptune would be discovered in this time. Many kinds of drama would appear to foreshadow future revolutions in physics in this time. Most notably the Michelson-Morley Experiment would take place foreshadowing the advent of relativity. Inventors like Edison and Tesla would flourish fashioning electricity into a modern technology. The Industrial Revolution would appear decisively.
- Early 20th century. In this time, thinkers like Planck and Einstein would appear to revolutionize physics. Modern physics would appear in this time transformed by notions like the atom, Big Bang, relativity, quantum, and other issues. All manner of physics issue would appear in this time shaping even

ATOMS (ATOMOS OR SMALLEST BIT)

now. Epic thinkers in this time would be Einstein, Planck, Bohr, Heisenberg, Feynman, Wheeler, DeBroglie, Compton, Rutherford, and many others. The Nobel Prize for Physics would appear. Particle physics, cosmic rays, the atom bomb, antimatter, the Big Bang, time dilation, the atomic nucleus, and other issues would appear.

- Middle 20th century. In this time going from the 1940s to 1960s, history would see World War 2. In this time, the atom would take center stage as a revolutionary drama. It would see developments like the nuclear reactor, atom bomb, H bomb, and uses of nuclear power. It would also see the Space Age, Information Age, Television Age, and other dramas appear. Einstein would die in 1955. He would spend his life searching for the grand unified theory of which later physicists would try to discover. Relativity would gain acceptance and epic inventions like the transistor, computer, laser, maser, and other things appeared.
- Late 20th century. After the year 1950, physics would mature from its revolutionary dramas of the early 20th century period. The Big Bang would gain acceptance in cosmology and nuclear physics would result in reactors and atom bombs. All manner of new issue would appear as well. The 21st century would take this era as its age and this continues on. Physics would gain issues like inflation, the cosmic acceleration, condensates, the search for the grand unified theory, quantum gravity, superconductivity, superfluids, neutrinos, quarks, black holes, pulsars, nonlocality, and the cosmic microwave background radiation. Major thinkers of this era would be Steven Hawking, Ed Witten, Nobel laureates, and other thinkers.

In all physics knows many eras and worldviews. The activity of physics in spirit is to try to understand the material world (universe). Before physics appeared, ancient peoples could only resort to superstition, religion, philosophy, and myths on how to explain the material world. Where these movements would fade or die, physics would arise to replace them.

Physics today is a vast labyrinth of concepts, beliefs, and sub-sciences. It has grown from its primitive ideas in Greece to become one of the major sciences. It is now an epic science with many complicated issues, theories, research paths, anomalies, characters, journals, and so much more. Physics continues to evolve, change, and transform now into a more complicated science. In the future other thinkers will expand on its teachings and issues further. This concludes our overview of the science of physics. In the next chapter, we will examine the nature of various ideas of physics needed for a continued education in science overall.

> "Physics had its seed in Greek philosophy. Greek philosophy was an attempt both primordial and primitive to explain natural things in terms of elements and reason. It would arrive at wrong conclusions and beliefs and this would affect science teaching for ages. In time, Man corrected for the mistakes of the Greek philosophers and in so doing invented the science of physics."
>
> -physics lecture

Chapter 3
Physics Concepts

"A concept is a creation of mind, an action of thinking. It has existence and meaning only in mind only with those who use intelligence."

-physics lecture

Fundamental Concepts

The Seven Fundamental Concepts

There are in physics, seven words that represent physics' most fundamental ideas. Their origin goes back ages and they are vital to the understanding of many sciences. They have perplexed thinkers for centuries with their seeming ease, intricacies, and surprises. Even now, Man's view of them is changing and it is unknown where they will go in their final vision. For the time being, they are at the forefront of today's physics discussion with more surprises being proposed about them. The seven basic concepts incidentally are mass, energy, massergy, space, time, spacetime, and mind. For now, we will explore our first fundamental concept, it has to do with expanse and infinity and is everywhere in the universe.

Space

"Space seems infinite and everywhere. Knowing what it is has been the scourge of physics eversince."

-Voigt

THE GRAND SCIENCE OF PHYSICS

The Concept of Space

Space. It is a word conveying notions of infinity, distance, expanse, and omnipresence. It is physics' most basic idea and is vital in the discussion of motion, speed, and force. Its nature has been a mystery for ages and it is still being explored. Space factors in such domains as chess, games, design, geometry, and many other subjects. In this section, we will explore the nature of space.

The Nature of Space (What is space?)

Space is fantastic word. It is so ordinary, yet it is so little understood. Space is deeply studied in geometry, the mathematics of shape and form. Qualities space is known to have are the following:

- Existence (it is real) –Emptiness (nothingness) –Infinity (or endlessness) –Dimension (three axes or lines at a right angle to each other) –Omnipresence (or presence "everywhere") –"Flatness" (or a mathematical rigidity) –No atomic nature (that is there is no such thing as an "atom" of space like matter) –Composed of points (objects of zero dimension) –Has the qualities of length, breadth, and width –Measurable qualities –Can be graphed (or depicted on a sheet of paper as a graph) –Allows for the creation of shapes and forms (like circles) –Is colorless or black of color –Allows for the passage of energies and material bodies –It is a unity (that is it is not divisible into smaller qualities) –Has no magical or divine nature –Has a conceptual quality –Is not thought to be a creation of mind –It does not "move", but allows for the measurement of motion of material bodies –Has a quality called direction (directions are like up, down, North, South, etc.) –It seems to have an intuitional nature (that is people "know" about space distances in some way) –It does not generate time, energy, or matter –It is not intelligent –It does not generate forces –It is a phenomena –It had a beginning in the Big Bang (the birth of the universe) –There exists a unique, shortest distance –It allows for structure –Paths through space are called "geodesics" –It is

FUNDAMENTAL CONCEPTS

a continuum (or continues) –It is identified with the vacuum –It can be investigated mathematically through geometry –It is fundamental to any discussion of the nature of the universe –It can interact with time as spacetime –Its smallest measurable unit is termed the "Planck length" –It can be "conceptualized" with math constructions called coordinate systems –Dramas that occur in space are called "events" -Vastness in space is referred to as "size" (Size is the idea of how big or small something is) -A "location" in space is referred to as a "place" (or an understanding of "where" something is) –Every question as to "where" something is refers to its "location" in space (place in space) –Technologies like spaceships can be invented to allow travel through space –Space is thought to be absolute, flat, and Euclidean (named for the Greek writer, Euclid who wrote the famous book, The Elements).

Ideas of Space

Space has been thought about by many thinkers. It has inspired many ideas found in surveying, geometry, and many domains. Basic ideas of space are:

- length –area –size –shape –form –volume –surface area –arc length –base –hypotenuse –coordinate –depth –width –circumference –perimeter –angle –symmetry –similarity –line fragment –ray –parallel lines –skew lines –straightness –bent –curvature –capacity -betweenness -fold

Location, Place (Where)

Space allows for the identification of place or location. A place is a point in space and implies where something is located.

Direction

Direction refers to space having "paths" called up-down, North-South, East-West, forwards-backwards, left-right, and so on. It is a quality derived from graphs and coordinate lines.

Emptiness (Nothingness)

This is space's most fundamental quality. It is said to be nothingness, the opposite of matter, the opposite of substance and fullness, the lack of the presence of atoms and energies, and so on. Emptiness and its nature have long fascinated thinkers. It continues on in discussions on the nature of space, nothingness, zero, and so on.

Omnipresence (Being Everywhere)

Space is thought to be everywhere. It is by the existence of space that an idea of location or where something is is defined.

Shape and Form

Space allows for the creation of objects called shapes and forms. Examples are points, lines, planes, circles, ellipses, parabolas, hyperbolas, squares, cubes, spheres, and so on.

Changelessness

Space is thought to be immovable, unchanging, and "still". Although its thought to warp and expand along with the universe, it is largely considered changeless.

Formlessness

Space as a reality is thought to be infinite, omnipresent, without limit, and so on. It allows for the creation of shapes (geometric constructions like circles) and forms (curves, surfaces, manifolds, anomalous formations), it however does not have a shape or form itself.

Continuum

A continuum is a "reality that continues". It refers to "realities" that seem infinite of extent or extend for great distances. Types of continuum are space, water (as in the ocean), rock, air, lava, gas atmospheres, and so on. Continuums have qualities like extent, size, depth, continuity, and vastness. Its thought in fantasy that there may exist "continuums" like ether, mind, spirit, soul, God substance, potential reality, etc.

FUNDAMENTAL CONCEPTS

Size

Space is said to contain all things and each thing is said to have a "size" (or amount of space it fills). Size can refer to things that are large and small. Its thought there is no end to largeness as its believed infinitely large things can exist. Its thought smallness can be infinite too (the infinitesimal), however no infinitely small thing is known to exist (a point is questionable however). Size is a basic quality of space and presented here is a list of things illustrating size.

- Medium size: people, rocks, mountains, hills, valleys, seas, meteors, animals, plants, trees, buildings, roads, dams, bridges, cars, houses, ditches, rivers, and the ocean.
- Small size: cells, bacteria, viruses, molecules, DNA, atoms, nuclei, proton, neutron, electron, photon, neutrino, quark, gluon, electromagnetic wave, and quanta.
- Large size: the Earth, the Sun, the Moon, planets, stars, comets, asteroids, galaxies, the Great Wall, the Great Attractor, the Local Group, and the universe.
- The smallest thing. Presently, its thought the electron, photon, neutrino, and point charge are the smallest things.
- Possible smallest thing. Its thought the smalleston (the smallest particle ever) or the infinitesimon (infinitely smallest particle) may be the smallest thing. They are not known to exist.
- The largest thing. Presently, its thought galaxies, the Great Attractor, the Sloan Great Wall, and the universe are the largest things.
- Possible largest thing. Its thought a multiverse (or reality containing all universes) may be the largest thing. It is not known to exist.

Dimension

Space is famously known to have three dimensions that being length, breadth, and width. Dimension can be thought of as a line. A line can extend infinitely in two directions. Now if you add another line at a right angle (ninety degrees) to that line, you would create a two dimensional space. Adding another line to the 2-D space turns it into a three dimensional space. Space can be denoted mathematically by the expression (x, y, z). A basic zero dimensional space is a point much like the tip of a pencil. A typical one dimensional space is then a line. A two dimensional space is a plane, a flat piece of paper, a countertop, or other "flat surface". A three dimensional space is a room, a volume, a box, or other such form. No one seems to know why space has the quality of dimension, however its discussion is fundamental to today's physics.

Simplex

For each kind of dimension, there is a "geometric object" that illustrates its nature or represents what it "means". A "simplex" then is a geometric object that illustrates what the kinds of "dimensional space" are like. Examples are the following:
- Point (zero dimensional space) –Line (one dimension) –Plane (two dimension) –Cube (three dimension) –Tesseract (four dimension) –"Kaluzact or pentact" (fifth dimension)

"Its thought space has more dimensions than three. They are curled or hidden somehow. No one knows if they exist, but finding them would have profound consequences."

-Voigt

Flatness

Space is said to have this quality. An example is a "flat" piece of paper. So long as there are no holes, creases, or crinkles, space is said to maintain its flatness. Flatness is identified with an ancient Greek philosopher named Euclid. He wrote a celebrated book called The Elements discussing his views on space. For centuries, space has been thought to have a basic flat nature that is it did not have the ability to bend or crease. For this space has been referred to as Euclidean or flat space. Today, there is a deep discussion going on that perhaps space has extra dimensions or that it can be "warped" or altered in some way.

FUNDAMENTAL CONCEPTS

Origin of Space

Its thought space had an origin or beginning in space with the Big Bang. It is unknown how space could be created in the Big Bang, but its discussion tends to be fascinating and filled with detail.

Geometry

Geometry is the mathematics of space. It is a subject taught in high school and deals with such things as circles, lines, triangles, and other such forms. It is a means to study the nature of space. An education in this subject is vital to learning about space. Basic "objects" known in geometry are the following:

- Point (zero dimension) –Line (one dimension) –Plane (two dimension) –Volume (three dimension) –circle –ellipse –parabola –hyperbola –square –ray –cube –parallelogram –rhombus –triangle –parallelipiped –pyramid –rhomboid –cycloid –astroid –cone –cylinder –rectangle –pentagon –hexagon –octagon –heptagon –nonagon -trapezoid -dodecahedron -sphere -helix -waveform

Please study books on geometry for a deeper understanding of this subject.

Line

A line is a "construction of space" consisting of points arranged in a "straight path". Lines are heavily studied in geometry. They are said to consist of points, be straight, have only one dimension, continue on forever, and be divisible into "line fragments". Famous kinds of "line" are called the ray, tangent, chord, radius, diameter, parallel line, and so on. Lines can "distort" to become sine waves, cycloid lines, curves, zigzags, and so on. Lines are fascinating issues to discuss in geometry and physics.

Plane

A plane is a "construction of space" consisting of an infinity of points in two dimensional "flat" space. A plane consists of points, is flat, continues on forever, and allows for the construction of shapes called circles, ellipses, squares, and so on. A plane shares much of the nature as a line. When a plane "distorts", it can wave, crease, puncture, have holes, and do other things. An analogy as to what a plane is a piece of paper, blanket, sheet, rag, or similar thing.

Volume

Volume refers to constructions of space in three dimensions. Examples are boxes, spheres, ellipsoids, paraboloids, cubes, hyperboloids, and so on. A volume shares a nature similar to lines and planes, but it is of three dimensions. A crystal lattice is a structure of volume.

Circle

The circle is one of geometry's most curious shapes. It is known for being round, a collection of points the same distant from its center, and other qualities. It was thought by the ancients as being the most perfect shape and the planets moved in circle-like (circular) paths. Its known that many masses "spin" or move in circles and this is intensely studied in mechanics. The circle has gone on to be geometry's most classic shape having inspired trigonometry, games, arenas, and many constructions.

Sphere

A sphere is a circle in three dimensions. Like the circle it is known to be round, have a center, and is practically identical to a circle in many ways. Its known that astronomical objects tend to appear in the shape of a sphere or like a sphere (oblate spheroid).

Conic Sections

A cone is a geometric object with a point top and circular base. A dunce cap, ice cream cone, blow horn, and similar object are examples of a cone. When a cone is "sliced", it produces curves and shapes that have been studied for ages. Examples are the circle, ellipse (distorted circle), hyperbola, and parabola. Conic sections in three dimensions are

FUNDAMENTAL CONCEPTS

called spheres, ellipsoids, hyperboloids, and paraboloids. Collectively they are called conic sections.

Curves

Curves are lineforms that behave like circles in a way. They tend to twist, bend, and be "rounded" much like a circleform. Types of curve are the helix, cycloid, parabola, hyperbola, astroid, and others.

Euclid's Elements

In ancient Greece, there lived a philosopher named Euclid. He wrote a textbook on geometry called The Elements that has become one of history's most famous books. For over 2000 years, it was regarded as the "classic" of the subject having influenced the Middle Ages, the Scientific Revolution, Rome, and other ages of Man. In its pages are discussions on geometry that would become basic teaching for all that time especially the section on Euclid's postulates. Please take time to read over this book as it influenced civilization so immensely even up till the present time.

Coordinates

Coordinates are "devices" used in geometry to locate a point in space. There exist many "systems" called the coordinate line, Cartesian (of Descartes), polar, spherical, hyperbolic, and so on. Please study their issues in analytic geometry.

Curious Objects of Space

There exist discussions in geometry on "shapes", forms, and spaces that have intrigued thinkers for ages. They go by names like the Moebius strip, Klein bottle, Calabi-Yau space, hyperspace, wormhole, tesseract, and so on. Please explore their issues.

Models of Space

Space is an intense issue to think about. This has lead thinkers to create analogies or "models" of how to think about space or explore its

issues. Examples of analogies are the following:
- Jello –A blank piece of paper –A blanket –A rubber sheet –A trampoline –A cloth sheet

The Void or Vacuum

Space is thought to be the "stage" of the universe, the backdrop by which all things happen in the universe. It is the "canvas" by which stars, planets, and material bodies exist and move. It is thought space itself is without substance or mass quality of some sort. Space is thought to be perfectly empty and is "simply" a barren emptiness extending everywhere into infinity.

Absolute Space

Isaac Newton thought of space as absolute. That is he thought it was rigid, changeless, possessing an existence apart from time, mass, and energy. He thought mass bodies moved through space, but did not affect the nature of space. This view of space endured for centuries until the coming of Einstein.

Relative Space

In the 20th century, Albert Einstein came along and proposed the idea of spacetime or space and time as one entity. In Newton's thinking, space and time are separate entities, but this notion was overthrown in favor of Einstein's in his relativity theories. Today, its thought space's nature can be influenced by time and mass bodies and this would be a radical notion to a physicist of Newton's era.

Views/Models of Space

Space has bewildered thinkers for ages as to how to "think" about it. Space has been "conceived" of as being "modeled" according to two basic visions:
- Newton's model. This is a vision where space is like a table top, a building framework, an immovable structure, and so on. Masses like balls can be placed on a tabletop, but the table is

FUNDAMENTAL CONCEPTS

not affected by the presence of a mass. In this vision, space and mass are distinct and separate. Mass moves through space, but does not affect the nature of space. Space is termed absolute.

- Einstein's model. In this vision, space is compared to a trampoline or rubber sheet. If a mass like a bowling ball is placed on the trampoline, the trampoline will bend or curve to accommodate the presence of mass (the bowling ball). In this model, space and mass are not distinct. Instead mass influences space to bend or curve and bending space influences what shape mass can take. Here space and mass influence one another and are different from Newton's model.

Technology and Tools of Space

Space is a fundamental notion to physics. So much so that thinkers have conceived all sorts of tools and inventions to use, measure, or exploit space in some way. Examples are the ruler, yardstick, meterstick, vacuum cleaner, surveying equipment, box, triple axis, slide rule, compass, T square, and other devices.

Grid

Take a flat piece of white paper. Place parallel lines on it evenly spaced going up and down and left and right. What is produced is a grid or sheet of graph paper. A grid is a "space construction" to locate points, establish an idea of distance, and construct geometric forms. It is a basic tool found in math and physics.

> "Space is not just nothingness. There is more to it than anyone knows or perhaps anyone can know. For so simple of thing, it could be the most complex thing of all."
>
> -class comment

Views of Space

Space has gone through many views in its history. Changes in view have at times been thought to be "revolutionary" and curious issues in physics. Presented here are discussions on some views of space through the ages.

- Superstitious space. Space was thought to be a magic, an extension of God's power, a magic controllable by the gods, and so on.
- God space. It was thought God created space by "His magic".
- Aristotle's space. Space was thought to be an extension of the fundamental substances of Earth, Air, Fire, and Water.
- Euclid's space. Space was an impersonal reality that allowed for geometric constructs like circles, squares, and shapes to exist.
- Tao space. Space was a creation of the Tao or impersonal reality of the universe in ancient China.
- Ether space. Space was thought born of the "ether", a fundamental substance.
- Atomist space. This refers to beliefs that space is composed of "atoms of emptiness".
- Newton's space. Space was an absolute domain independent of mass or energy, it was a "stage" to allow mass objects to move through.
- Mach's space. Ernest Mach, a teacher of Einstein thought space did not exist or was an extrapolation of cosmic masses or "totality".
- Einstein's space. Einstein thought space was a union with time as spacetime. He thought space could "curve", be disturbed by mass, and perhaps cause gravity.
- Mind space. This is a controversial notion out of philosophy that space is a creation of mind.
- Particle space. This is also a controversial notion that space may have an "atomic nature" and its particle is called the "emptyon".
- Speculated Space. There are discussions that space has more dimensions than three, perhaps four, five, ten, 26, or how many and hidden dimensions as well.

FUNDAMENTAL CONCEPTS

The Frontier of Space

Space today is a great topic of discussion of research in physics and science fiction. Dreamers and scientists discuss all sorts of fantastic ideas of space and wonder if their musings have any reality. Compiled here is a list of ideas and discussions about space currently the "fashion" in science.

- Space may have "extra dimensions". This refers to notions that space may have more than three dimensions. It may have four (hyperspace), five (Kaluza space), infinite dimensions (Hilbert space), fractional dimensions, curled (or unseen) dimensions, negative dimensions, and so on. So far these notions are unconfirmed, but efforts are underway to explore them.
- Space may have a wave or particle nature. A particle of space is "called" the "emptyon" or "empty particle".
- Space may have an "edge", "holes", rotates, contracts, expands, be created by a quantum fluctuation, or has a bubble-like or foam nature.
- Space may be able to "end" in a black hole or it may "decay".
- A "wormhole" is the idea of a "bridge" through space.
- Hyperspace travel through space may be possible.
- Space may have an opposite in an idea called "antispace" or "unspace".
- Space may be able to tear, crease, rip, puncture, fold, warp, distort, bubble, crinkle, twist, wave, crumple, or change in some way.
- Space can be created or destroyed by magic.
- Spaces of two dimensions (called Flatland) may exist. This leads to the notions of universes of zero, one, two, or more than three dimensions existing.
- Space can unite with time as spacetime. Space can interact with mass, energy, or time in physical interactions producing phenomena. Examples are space-mass interactions, space-energy interactions, and so on.

THE GRAND SCIENCE OF PHYSICS

- A "bulk" is the idea of a higher dimensional space that somehow leads to space. It is discussed in advanced physics in a subject called M theory.
- Space can generate forces or energies. The idea of a space energy or space force ("emptity") is thought to be possible.
- Space is somehow thought to be born in the Big Bang and shares a common nature with mass, energy, and time. Space may be born of a "unified substance" called MEST or "mass-energy-space-time substance".
- Space does not exist, it is an illusion. A universe where space does not exist is referred to as the infinite "one place" universe (Infinite Here).
- Speculated kinds of particle of space are called the "emptyon", voidon, emptino, semptyon, emptychronon, emptychronino, semptychronon, etc.
- Its thought space contains violent displays of energy that can create subatomic particles in pairs (both matter and antimatter versions).
- Its thought space could be born from a "quantum fluctuation".
- Its thought space may cease to exist in an event called the Big Crunch.
- Superspace and subspace are ideas of "kinds of space" somehow above and below the "normal continuum of space". Science fiction writers have taken to using them to allow hyperspace travel.
- Speculated kinds of space are called paraspace, quasispace, hyperspace, antispace, prespace, postspace, spirit space, mind space, God space, superspace, subspace, ultraspace, ether space, negative space, and so on.

Philosophy of Space

This is a subject that deals with the discussion of the nature of space. It is found in philosophy in many works.

FUNDAMENTAL CONCEPTS

Perception of Space

This refers to how thinkers are "aware" or conceive of space. Notions include ideas of location, place, order of things, and so on.

"Its known that the shortest path between two points is a straight line, but what if there were a shorter path?"

-class discussion

Hyperspace

There is in science fiction the idea of a four (or multi-dimensional) dimensional space "transcendental" to "ordinary 3-D space" called "hyperspace". Hyperspace is used by aliens to travel fast across a galaxy or around the universe. It is a favorite theme among science fiction writers to make for "easy star travel" in "economical time" as found in the Star Wars movies. Beliefs in hyperspace circulate in modern physics in the form of hidden dimensions of space, wormholes, black hole tunnels, and so on. There is no evidence it exists. However if it is someday discovered and can be controlled, its thought it would have revolutionary implications for future space travel. For the time being, it is a curious notion and one of physics' more outlandish ideas.

False Vacuum

Alan Guth, a cosmologist proposed famously that the universe underwent an era of "super expansion" called "inflation" after the Big Bang. An idea was created that perhaps the universe was born from a "reality" called a "false vacuum". Although the false vacuum is not known to exist, its thought to be a "state of space" of some kind. Today, it's a controversial notion whose nature is unknown.

Spacewarp (Hole or Rip in Space)

Its thought regions of space can "warp" or become "unusual" of their nature. Inside a spacewarp, travel in time or to exotic locales or universes is thought to be possible. Its thought a black hole would allow for the creation of a spacewarp as the laws of physics breakdown here. For the time being, its not known if any spacewarps exist or are possible.

Length Contraction

Its known in special relativity that when an "observer" travels near the speed of light that space will contract. This means that space seems to grow "smaller" or that distances seem not to be vast.

Topology (Topography)

This is a branch of mathematics dealing with space, surfaces, "manifolds", and issues related to space. It explores the nature of surfaces of rubber, a tabletop, a piece of paper, and so on. It discusses such issues like tangrams, the Moebius strip (piece of paper with only one side), the Klein bottle, and so on. Its described as a very difficult subject and can be explored in other works.

Geography, Geodesy

These are subjects that deal with making maps and finding elevation. They involve the concept of space immensely and are sciences derived from space.

Space Travel

Space travel refers to technology able to travel in space like rockets, space probes, solar sails, star ships, and so on. Science fiction writers like to invent all kinds of ways to travel in space. Today, dreamers are hard at work devising ever more outlandish ways to travel in space. Astronauts who do travel in space know space is a very dangerous environment. They know direct exposure to space can be deadly and they must make certain all their technology works right.

Distance in Space

Its known that the distances between stars are astronomically great. To build a starship able to travel to another star would take thousands or millions of years. Its thought a starship would require hundreds of generations of people to live and die aboard such a ship to complete such a journey. For now, it is considered a domain of science fiction as to whether "star travel" will possible someday or not.

Geodesic Dome

The famous thinker Buckminister Fuller conceived a dome made of interlocking shapes. It has since gone on to be a fascinating object in science used in making houses, playground objects, and other buildings.

Space has always been a mysterious idea in physics. It represents the opposite of matter, fullness, and substance, yet it baffles as to what nothingness is. It is today one of physics' more compelling issues and interest on it continues on.

In the next section, we will explore a discussion on the sister concept of space, time.

> "Try to grab hold of space with your hand. You realize you have nothing and this is what the pursuit of space has been."
>
> -class comment

Time

The Concept of Time

Time is like space just as fundamental and fascinating. It too is an ancient notion and factors in virtually all cultures and worldviews. Time's nature is currently discussed with many intriguing notions in physics. Presented here is a discussion on this most famous issue.

> "The only reason for time is so that everything does not happen at once."
>
> -Albert Einstein

The Nature of Time

Time is a perplexing issue. It too has a deep intricacy and requires an intense study. Time is thought to have these qualities to its nature:

- It flows from the past to the future passing through the present moment –Time cannot flow from future to the past –Time is fundamental to the universe –There exists such a thing as the arrow of time and physics recognizes various kinds like the cosmological, radiative, the weak, and entropic arrows of time –Time can "dilate" or change as a person or thing travels closer to the speed of light –Time is not an intelligence or born of consciousness –Time is not motion, but is vital to measuring moving things –Time does not generate space, mass, energies, or forces –It exists –Its flow is regularity and does not change –Time has no magical or divine nature –Time can be measured by clocks –Causality is the notion of causes precede effects and is intertwined with the arrow of time –Time can be conceptualized –Time is a phenomena –Time is coexistent everywhere along with space –Time had a beginning –The flow of time is called duration –Time is not a living thing –Time has no atomic nature (that is there is no such thing as a particle of time) –Time is not supernatural –Time cannot be created nor destroyed –The smallest measurable quantity of time is called the "Planck time" –Time has no dimensional quality, but is thought to be a dimension of space as spacetime –Constructions for measuring time have been created and examples are the notions of age, history, second, year, cosmic time, geologic time, and so on.

Perception of Time

This refers to an awareness of time or what time it is. Its known that every person has in some way a perception of time.

Time's Three Qualities

Time is known to come in three distinct qualities called the past, present, and future. Discussion on these qualities are:

FUNDAMENTAL CONCEPTS

- Past. This refers to all events that have occurred or "passed" by. The past is remembered through history.
- Now, Present. This refers to events happening in an instant of time called the present. The present is studied in current events.
- Future. This refers to events to occur in a later time. It is studied in futurology, envisioning, and prediction. A subject called prophecy claims to know the future by magical means and is considered suspicious in physics discussion. Various suspect subjects that claim to be able to know the future go by the names of astrology, Tarot, fortune telling, prophecy, precognition, numerology, divination, and so on.
- The arrow of time represents time's flow from past to future. It is not possible to time travel.

Time is often compared to a river to explore its nature.

Duration (Arrow of Time)

This refers to the flow of time, of time passing from future to past through the now or present.

Location in Time (When)

Events are said to happen according to some time. This establishes an idea of "when" something occurred.

Smallest Measurable Time

Time is thought to flow from future to past and has "instances" called "moments". Scientists today can measure time at "intervals" of a second and nanosecond. It is unknown what the smallest unit of time is (the quickest moment), but efforts are underway to measure time at quicker intervals. The Planck Time is a discussion of an "instant in time" called 10^{-43} second and is found in cosmology.

Temporal Finitism

This refers to beliefs that time must be "finite" or somehow end. It discusses how time may have begun and the "eventuality" that it will end. It is more a topic of philosophy and has a deep discussion.

The Measurement of Time (Horology) (Timekeeping Devices)

Time is a notion deeply vital to everyday life. Various inventions and devices have been created to measure and categorize time. Many of these inventions are ordinary parts of life and many cultures have their own unique tools and inventions of time. Examples of such devices are:

- Clocks (Horologes). These are devices use to measure time in quantities called seconds or hours. Examples are grandfather clocks, wristwatches, wall clocks, electronic clocks, atomic clocks, chronometers, hourglasses, gnomons, tower clocks, alarm clocks, cuckoo clocks, pendulum clocks, chess clocks, natural clocks, and so on. The study of clocks is called horology or clock science.
- Calendars. These are devices for measuring the flow of time in quantities called days, weeks, months, and years. Calendars are ancient inventions and virtually all ancient societies had their unique version. A year is defined as the time it takes for the Earth to travel about the Sun and comes in a time span of 365 days. Calendars come in many varieties and examples are the Julian, Gregorian, Hebrew, Chinese, Indian, Islamic, Aztec, and so on. Special days have been created to be placed on calendars. Referred to as holidays or "holy days", they tend to be days set aside to celebrate some feature of life. Examples are Christmas, Halloween, Easter, a birthday, Thanksgiving, Kwanzaa, Hanukkah, and so on.

"A calendar is a system for ordering time by making the world seem ordinary and comprehensible as day passes to day."

-class discussion

FUNDAMENTAL CONCEPTS

Calendar (Calendrics)

Many societies have created a "device of time" used to measure time, define seasons or years, and be as a standard by which civilization moves. Called calendars, they are epic constructions dating from the earliest civilizations. A brief discussion on the modern calendar goes as follows:

- There are 12 divisions called "months". A month is the time it takes for the Moon to orbit the Earth. Each month is given a name like January (Janus), February, March, April, May, June, July (Julius Caesar), August (Augustus Caesar), September (seven), October (eight), November (nine), and December (ten).
- The calendar unit called the "year" is the time it takes for the Earth to go about the Sun. It is divided into units called weeks.
- The week is divided into seven units called "days". A day is the time it takes the Sun to go from morning to night and lasts 24 hours. Days go by names like Monday (Moon's day), Tuesday (Tiw's day), Wednesday (Wodin's day), Thursday (Thor's day), Friday (Freyr's day), Saturday (Saturn's day), and Sunday (Sun's day).
- A day is divided into 24 hours with an hour lasting 60 minutes. A minute will last 60 seconds.
- Days are sometimes set aside as "holy days" or holidays, times of celebration. Famous ones are Christmas, Halloween, Thanksgiving, Columbus Day, Easter, and so on. Holidays will commemorate religious events, people, US presidents, independence, a nation, the queen or king's birthday, an ethnic time, or some other grand event.
- There exist curious "times" called the leap year and leap day.
- New Year's Day is January 1.
- A year can be divided into "seasons" or times of weather and agriculture called fall, winter, spring, rainy, dry, monsoon, and summer.

Julian and Gregorian Calendars

In ancient Rome, there was a calendar system inherited from the ancient society of Babylon. It was found by the Romans that this ancient calendar was mistaken in many ways. The famous Roman general Julius Caesar ordered a revision to this calendar adding two extra months named July (Julius Caesar) and August (Augustus Caesar). This corrected calendar would be the standard of Europe until the time of a Roman Catholic pope named Gregory. In his time, the Julian calendar was found to be flawed and needed correction. This pope ordered a revision to the Julian calendar and this came to be called the Gregorian calendar. Today, the Gregorian calendar is the standard calendar for Europe, America, and much of the world.

"In Caesar's Rome it was found that the calendar was broken and it needing fixing. Centuries later, this fixed calendar was found to be broken again and needing fixing. Its thought centuries later the calendar will be broke again and it will require another fixing yet again."

-class discussion

World Time

This refers to attempts to establish a "uniform system" of time around the world. Various offshoots of this are Coordinated Universal Time, Greenwich Mean Time, time zones, sidereal time, and so on.

Sundial

The ancients were baffled by time and wanted to measure it more. They studied the Sun and its motions intensely. Some ancient thinker then proceeded to devise one of the earliest kinds of clock that would be called the "sundial". The sundial is a circle with a stick on it. As the Sun moves in the sky, it casts a shadow from the stick. This shadow then moves about the circle at a rate that the Sun moves. In this way, ancient thinkers realized that they could devise a "clock" by which to gage time from morning to night when the Sun goes down and the shadow disappears. The sundial has gone on to be one of history's more enduring inventions as it was a kind of predictable and trustworthy clock.

Hourglass

Ancient thinkers invented a "clock" consisting of two "glasses" connected by a small tube. One glass container contained an amount of sand by which the sand would "drip" into the lower glass container. This "dripping" of sand would last for about an hour and a device using this is called the "hourglass". No one knows when or where or with whom it was first invented, but it is a classic invention in the history of clocks.

Clepsydra

Also called a water clock, this is a time-keeping device where an amount of water drips according to some amount of time. It was used extensively in Ancient Egypt and other early civilizations.

Atomic Clock

In the 20th century, various physicists realized that radioactive elements could decay with very good precision. This inspired them to build clocks using radioactive elements also called atomic clocks. Atomic clocks are known for being highly accurate and some of the best clocks ever made.

Biological Clock

Its known that people, plants, and animals seem to have a sensitivity to time. It can also refer to where people age or move from state to state like child to adult. It is unknown how the biological clock operates, but it is a curious issue in time and biology.

Astronomical Clock

This is a modern drama in science to devise clocks attuned to astronomical events. Clocks have been devised based on pulsar energy releases, activities of the Sun, planet movements, and so on.

THE GRAND SCIENCE OF PHYSICS

Constructions of Time

Time has been measured by many cultures "through time". They have invented notions of how to define a "time unit" or time measurement. A collected list of such "time units" are the following:

- second (SI unit) –minute (60 seconds) –hour (60 minutes) –day (24 hours) –week (seven days) –month –year –leap year –decade –century –milennium –age –epoch –era –solar year –sidereal time –geologic era –"geologic time" –"cosmic time" –nanosecond

Time Models

Various thinkers have modeled time to seem like the following:

- a river –a line –a string –a piece of paper –a length of yarn –a gust of wind

Geologic Time

This is a construction of time born in the science of geology. Its thought the Earth is immensely old about 4.5 billion years old. Various thinkers have sub-divided the time of the Earth's existence into "ages". These ages are identified with ages of life, catastrophes, types of rock, and geologic events. Its thought the time of the Earth is divided into "eras" called Proterozoic, Paleozoic, Mesozoic, and Cenozoic. The Paleozoic is said to be an era of primitive plants, fishes, and amphibians. The Mesozoic is the age of the dinosaurs. The Cenozoic is the age of mammals, birds, and Man. These eras are sub-divided into other ages called Mississippian, Permian, Jurassic, Triassic, Cretaceous, Eocene, Miocene, Oligocene, Pleistocene, and Holocene. The Holocene is considered the age of Man. Please explore the discussions on geologic time in more definitive books.

Cosmic Time

This is a construction born of the science of cosmology. It refers to a timeline of the universe from the Big Bang (thought to be 13.7 billion years ago) to its end (whenever and however that will be). Its thought the universe has gone through ages called the Planck, inflation, cooling, and other eras. These eras are discussed in cosmology books.

Views of Time

Time's discussion has gone through many views through ages. Presented here is a discussion on various views on what time is thought to be.

- Superstitious time. Primitive people have thought time was a magic of God or the gods. They did not know what it was, but they knew it to be a flow of something.
- God's time. This is the idea that God creates or somehow "drives" time to flow.
- Aristotle's time. Time was thought to be a consequence of the fundamental substances of Earth, Air, Fire, and Water.
- Heraclitus' time. Time was a "flow", but of unknown nature.
- Indian time. Time was thought to be a mysterious flow born of the Brahman in Ancient India.
- Tao time. In Ancient China, it was thought time was born of the Tao (impersonal reality) or of the "interaction" of Yin and Yang.
- Ether time. Time was thought to be born of the "ether", a fundamental liquid reality.
- Kant's time. Immanuel Kant was a famous philosopher who maintained time was part of the "conceptual framework" of the universe. He maintained it was an idea or tool to think about and understand the universe.
- Newton's time. Time was a flow of unknown origin that existed alongside yet separate from space.
- Mach's time. Ernst Mach, a teacher of Einstein thought time did not exist or originated from cosmic masses.
- Einstein's time. Einstein thought time was a union with space as spacetime. Time could "dilate" (or change) as observers travel near the speed of light.

THE GRAND SCIENCE OF PHYSICS

- Mind time. This is a controversial notion that time is a creation of Mind.
- Nonexistent time. This is a controversial notion that time does not exist or is an illusion of the senses.
- Particle time (quantized time). This is a controversial notion that time has a particle nature and its particle is called the "chronon".
- Linear time. This is the idea that time moves in a line and only a line.
- Cyclic time. This is the idea that time moves in a circle repeating cycles.
- Science fiction time. This refers to controversial notions that time is generated from the "string" (a hypothesized subatomic entity), a brane, or a collision of universes.

The Frontier of Time

Time's discussion is filled with many stimulating notions that seem to be more like science fiction than physics. Presented here are notions currently discussed on time.

- Time can be reversed moving from future to the past. This is referred to as the reversed arrow of time, time reversal, or reversed causality.
- Time may have a dimensional nature, can "decay", or come to an "end".
- Time can have wave or particle qualities. A particle of time is called the "chronon", searches for it in particle accelerators have found nothing though.
- Time may have a magical nature, bubble qualities, or other bizarre qualities.
- Time may have a "chance" or chaotic nature, flowing at irregular intervals.
- Time can be "stopped" (timelessness), started, hurried, or slowed.
- A "crazy universe" is a universe where time flows chaotically.

- A universe where all time has stopped is called an "eternal now".
- Time is an illusion, it does not exist, or it is a creation of mind.
- Time may have an opposite in an idea called "antitime".
- Time may exist in a hyperspace universe in the idea of "hypertime".
- Time may come in versions called supertime, subtime, hypertime, antitime, ultratime, quasitime, pretertime, pretime, posttime, quantum time, and so on.

Time Dilation

Its known in relativity that as something travels faster approaching the speed of light, time will slow down or appear to slow down. Called time dilation, this effect has fascinated science fiction thinkers eversince. Its thought if a spaceship could be built that can travel at speeds near that of light, then on-board crew would experience time slowing down "relative" to slower objects. Such a space crew could go on a "trip" lasting only a few years in their "awareness", but time could pass by for literally thousands of years from where they came from. When the spaceship crew return home after their "brief" trip, they find they are thousands of years into the future.

Chronon (Particle of Time)

Chronos is the name of the ancient god father of the Greek gods and personification of time. His name has been used to name the particle "chronon" or particle of time. Various thinkers had believed time consisted of "atoms" or points and that these atoms could be discovered or used. Today, its not thought the chronon exists, but it is discussed in modern physics as of now. Other kinds of "particle time" are the chronino, emptychronon, schronon, chronetto, and so on.

Timewarp

Its thought time may have more of a nature that anyone thought. Its speculated regions of space may undergo "warpage" or become enigmatic astronomical objects. Such objects would allow for time travel or travel to "exotic places and times". While its not known that any timewarps exist, its thought inside black holes the laws of physics breakdown and thus a black hole would become a timewarp.

Beginning and End of Time

Its thought time had a beginning in the Big Bang or will have an end in the End of the Universe. No one knows how time began to flow, what originates time, and what influences time to have its nature. However, time has continued to be fascinating and today these issues are among some of its greatest mysteries.

Eternity and Eternal Now

Eternity is the idea of infinitely long lasting time. The Eternal Now is the idea of a reality where past, present, and future all "exist" and happen at "once".

Hypertime

This is an idea analogous to hyperspace. Here time can from from one time to another time "superfast".

Time Travel

In various science fiction stories, probably the most famous is HG Wells' classic The Time Machine, various "heroes" use a time machine to travel to various eras of history. This had lead writers to create stories of people traveling to the age of World War 2, Napoleon's time, and various eras of history. While time travel is a fascinating subject, it is currently not believed to be a reality. It is thought that if it could be real, it would violate a discussion called the Grandfather Paradox. Briefly stated, it says that if you could travel into the past, you could kill your grandfather or some other ancestor. If you did that then this would start a chain of events that would lead to the prevention of your birth, thus you could not exist to do such an act. This has usually been the foremost argument that time travel is impossible. However, in today's physics, there are discussions that perhaps this argument is flawed

FUNDAMENTAL CONCEPTS

and that time travel is indeed possible. The discussion on this issue tends to be intense and technical, the reader is encouraged to explore stories and books using and discussing this issue.

Time Travel Fantasy

Time travel is a fascinating issue in science. Many thinkers have taken to writing works or using it in some way. They have created "heroes" who travel back in time to stop a villain, change the future, or alter history in some way. It is a source of many stories and famous works employing time travel are:

- Time Machine (HG Wells) –Star Trek –Quantum Leap –Dr. Who –Battlestar Galactica –Buck Rogers in the 25th Century –Time Bandits –Arthurian scifi fantasy

Future Travel

This is a kind of time travel where a person in the present can "travel" to a future many thousands or millions of years into the future (in a second). There exist many time travel stories that use this theme. However, people who do future travel cannot travel back to their own time as time travel to the past is impossible.

Father Time, Chronos, and Kairos

This is an imaginary being said to be the personification, god, lord, or embodiment of time. He is a purely fictional character and is used in many stories. Many societies maintain some kind of "divinity" of time, fate, the past, and the future. While its known none of these entities exist, they are a curious part of world mythology.

Wheel of Time

Many societies maintain time repeats in cycles or moves about like a wheel. It discusses that seasons repeat themselves, that the Earth goes about the Sun in a cycle, and that historical patterns repeat themselves. It is an issue that has become intertwined with religion, superstition, the occult, and other issues.

Time Capsules

These are objects or chambers set aside to be filled with publications, artifacts, and other curio to be buried somewhere. They are not intended to be opened except in some later time. Time capsules are fascinating things and tend to be projects of schools, colleges, or for some other "special" event. Many kinds of time capsule have been created, probably the most famous is the so-called Tomb of Civilization. Time capsules are a way to teach history and commemorate events in some way.

Please consider making a time capsule.

Absolute Time, Relative Time

Isaac Newton, the famous physicist believed that time was a like a river. It flows from past to the future, but it cannot be hurried or slowed and it is an entity separate and distinct from space. For centuries, this has been the prevailing view of time until the coming of Albert Einstein. Einstein proposed that time was a dimension of space, that time was joined with space as spacetime, and that time can be changed of its flow. These were powerful claims and virtual heresy in physics. Time along with space was thought to be "absolute" or rigid and unchangeable in some way.

"Absolute, true, and mathematical time, in and of itself and of its own nature, without reference to anything external, flows uniformly and by another name is called duration. Relative, apparent, and common time is any sensible and external measure (precise or imprecise) of duration by means of motion, such a measure – for example, an hour, a day, a month, a year – is commonly used instead of true time." -Isaac Newton (Principia)

Philosophy of Time

There is a branch of philosophy dealing with thinking about time. It discusses ideas like duration, the past, the future, what causes time to flow, and so on. It can be explored in many works found in libraries.

FUNDAMENTAL CONCEPTS

Time Management
This subject discusses how to use time in various activities of employment.

Unreality of Time
There exists in philosophy and physics an obscure discussion. It refers to beliefs that perhaps time or time's flow does not exist. Many writers have written books trying to prove time does not exist. They try to invoke that it is an illusion of senses, an illusion of mind, or an illusion of a deeper reality. For the time being, physics does not accept such notions as time has a clearly observable nature and is known to exist. Please read about this curious sub-issue in the issue of time.

Reversed Time
It is not known why time flows from future to past, but it is known that reversed time (flow from past to future) does not exist. It is thought to be forbidden by a discussion called the Grandfather Paradox and maintains time travel is impossible.

Timelessness
This refers to a "state" where time does not flow or that something seems forever. While its thought a "timeless state" does not exist, its thought some things are "timeless" as they seem relevant in any age.

Imaginary Time
This refers to an idea where time is thought to be a creation of thought or mind. It claims that time can be imagined, controlled by thought, or something else. For now, it is discussed in advanced physics, but is not known to be real.

Impermanence
This refers to things always changing or not being permanent of state (impermanent). It is discussed heavily in Buddhism and involves time influencing the change of state of things.

Knowing the Future

Many people around the world and through the ages claim to know the future. They are called by such words as prophets, seers, fortune tellers, intuitives, clairvoyants, mages, and so on. They claim to know knowledge of events to come like wars, assassinations, great events, births, deaths, and other dramas. Famous people who claimed this ability are called Nostradamus, St. Malachy, Mother Shipton, the Brahan Seer, Jeane Dixon, and many others. They are known to have given "prophecies" or accounts of what is to come in the future. Nostradamus himself is famous for writing a book called The Centuries outlining prophecies of events to come after his life in the 16th century. For the time being, modern physics does not believe in such claims and prophets. No claim of knowing the future has withstood analysis and today it is thought to be an issue of pseudoscience, superstition, and fraud. Issues associated with this are called precognition, retrocognition, clairvoyance, prophecy, fortune telling, etc.

The Future

The future refers to events to come, tomorrow, or a time that will be. The future is studied in a topic called futurology. Futurists are visionaries of the future discussing science fiction issues and the world to come. Types of "scientific future-telling" are weather forecasting, stock predictions, estimation, envisioning, and so on. Such activities do not involve reference to the psychic, magical, and pseudoscientific.

History (Chronology and Chronometry)

History is a science that studies events of the past. Its subject matter consists of wars, political movements, leaders, cultures, trends, civilizations, rises and falls, etc. It is a deeply complicated subject with branches like physics history, world history, US history, and so on. Famous characters out of history are Caesar, Christ, Buddha, Napoleon, Hitler, Einstein, JFK, Washington, Roosevelt, Robin Hood, and many others. It is studied at virtually all colleges and is found worldwide in libraries.

FUNDAMENTAL CONCEPTS

Alternative History

In science fiction, there is a literature that imagines stories where history proceeds in a way different from what the history book reads. This is known as alternative history. Here a writer may imagine Hitler wins World War 2 or Columbus does not discover America. Anything can be imagined and anything goes in these kinds of stories.

Time has always been a mysterious idea in physics. No one knows what makes it flow nor why it only goes from past to future. Its issue has bewildered thinkers as they don't know just what it is and today it hides many secrets still.

"Perhaps seers can know the future. They can then know about wars, acts of science, events of death, highlights, and tragedies yet to be. If the future could be known, a whole different world would be."

-Voigt

In the next section, we will now review a discussion of a concept different from space and time, it is an idea of their union as 'spacetime'.

Spacetime

The Concept of Spacetime

In physics, Albert Einstein and other thinkers proposed the fantastic notion that space and time were joined as one reality, called spacetime or timespace. Today, this idea is fundamental to the discussions of gravity, the Big Bang, and other subjects. Spacetime of itself is neither space nor time, but a blend of both. For this, it has taken its place in physics as its own fundamental concept and is the subject of this chapter.

The Nature of Spacetime

Spacetime is an intriguing notion. Its thought mass objects can "influence" spacetime to create gravity effects. It has an intricate nature

and ideas concerned with it are the following:
- Spacetime is not an intelligence –It does not generate mass, energies, or forces –It is not space nor time, but a blend of the two –It does not have a magical or divine nature –It can bend, warp, and curve –It is said to contain four dimensions (three space dimensions and time as a dimension) –It has no particle nature (hence there is no "emptychronon" or particle of spacetime) –It breaksdown of its nature in an "object" called a singularity (a point of infinite gravity, pressure, and density) –It is used in mathematical models –It is used in the construction of "radical" physics theories –The whole of spacetime is called Minkowski space –Events are points in spacetime –Spacetime is independent of any observer (or "mind" looking at it) –A worldline is a "path" through spacetime –A distance between any two points of spacetime is called an "interval" –Spacetime has kinds of symmetry (or opposite halves) like axially symmetric, spherically symmetric, static, and stationary –Spacetime had a beginning –Pairs of events in space time come in three categories of "light-like", "time-like", and "space-like".

Curvature of Spacetime

Its thought mass can cause spacetime to curve, bend, or warp. This effect causes gravity. Curved spacetime is studied in a subject called Riemannian Geometry, or the geometry of spacetime.

Spacetime Interval

This is a belief found in relativity. It is about spacetime having "distances" or intervals that somehow influence the nature of spacetime.

Worldline (Geodesic)

A worldline is a line through space, however a worldline is also a line through spacetime as well. It refers to the path an object travels in space or spacetime. Kinds of worldline are a trail, bike path, sidewalk, chalk line on a street, etc. A worldsheet is a space or plane that contains worldliness.

FUNDAMENTAL CONCEPTS

The Frontier of Spacetime

Spacetime currently occupies a foremost position at the forefront of physics research today. Many "radical" ideas have been proposed on its nature and compiled here are some musings:

- Spacetime had a beginning in the Big Bang.
- Spacetime may have an end in the Big Crunch.
- Spacetime may not exist.
- Spacetime has a wave or particle nature.
- Spacetime may have magical, divine, bubble-like, or chaotic qualities.
- Spacetime can radioactively decay or "end".
- Spacetime may come to an "end" in a black hole.
- Time travel in spacetime is possible.
- Spacetime may have an opposite in an idea called "antispacetime".
- Spacetime atoms may exist or a particle called the "emptychronon".
- Spacetime is created by God, is controllable by magic, and so on.
- Spacetime warps may exist.
- Virtual spacetime refers to a counterpart to real spacetime in a nonexistent way. It is thought to be the source of virtual particles.
- Spacetime has all the fanciful qualities of space and time in fantasy as well.
- It is known in relativity that length contraction and time dilation occur as one travels close to the speed of light. In a combined notion in spacetime, these ideas are combined to form the notion of "dilacontraction" or spacetime contraction and dilation all as one.
- Kinds of spacetime may exist like paraspacetime, quantum spacetime, hyper-spacetime, quasi-spacetime, ultra-spacetime, super-spacetime, and sub-spacetime.
- Spacetimes of dimension greater than four have been devised. Its thought extra dimensions of spacetime may curl or exist hidden unknown from physicists.

Thinkers of Spacetime

Spacetime tends to be one of physics' most difficult and enigmatic ideas. It is not easy to explain and requires a deep study of its treatises to realistically comprehend. Major thinkers into spacetime are the following and their books can be explored:

- Einstein –Ed Witten –Stephen Hawking –Roger Penrose –EA Poe –HG Wells –Hermann Minkowski –Hendrik Lorentz –Bernhard Riemann –Paul Ehrenfest –Immanuel Kant –Theodor Kaluza and Oscar Klein –John Wheeler –many thinkers and treatises

Origin of the Idea of Spacetime

The idea of spacetime existed long before Einstein came along. It has been a speculation of ages that perhaps space was joined with time as one reality. Its known the famous writer Edgar Allen Poe conceived of spacetime and the ancient Inca referred to spacetime as "pacha".

Models of Spacetime

Spacetime has been "analogized", conceived, and compared to such things as:

- A piece of paper with lines drawn on it –A trampoline –A rubber sheet –A river -Jello

The Philosophy of Spacetime

This refers to a branch of philosophy discussing spacetime by thinking. It is mostly an obscure subject, but the reader is encouraged to explore it.

Hidden Dimensions (Higher Dimensional Spacetime)

Theodor Kaluza, a physicist in Einstein's time proposed the notion that spacetime may have a fifth dimension. At the time, it was a radical notion and no one believed in it. However, later thinkers proposed the notion that spacetime may have more dimensions, but these dimensions remain hidden or are "curled" up away from detection by physicists. Today, physicists search for evidence that hidden dimensions of spacetime exist, but so far have found nothing. The discussion of this is found in such issues as supergravity, M theory, and other topics in advanced physics.

Bulk

There is discussion in advanced physics that perhaps spacetime of dimension greater than four may exist. This kind of spacetime is called "bulk". It is not known to exist, but its thought the universe may be born of a bulk or a "bulk universe" exists as a parallel universe.

TARDIS

In the Dr. Who science fiction TV show, there exists a kind of spaceship called the Tardis. Tardis is a word meaning "time and relative dimension in space". It is a spaceship that appears like a phone booth, but inside it it has the capacity of a "house". It is a fictional spaceship that's can travel anywhere in space and in time. It is used by a fictional "hero" called the Doctor and is a popular TV show in Britain. The Tardis illustrates various notions about spacetime in a science fiction format.

Spacetime is among some of physics' "newest" ideas. It is the union of space with time saying time is a dimension of space. Today, its discussion is topical in general relativity and related "theories" and research into it continues on.

> "Spacetime seems like a magical word. It conveys an unknown meaning that perhaps magic is possible in Nature after all."
>
> -Voigt

In the next section, we will review the idea of 'energy'.

THE GRAND SCIENCE OF PHYSICS

Energy

"Energy represents God's magic or divine power. It is power to use Nature, to make tools, and to show Man's mastery over natural forces."

-class comment

In this coming section, we will examine the nature of an idea so vital to physics and the modern world that science could not be without it. Called energy, it had its origin in the discussion on the nature of magic. Today, its found everywhere from forces to engines and is vital in so many ways.

The Concept of Energy ("The ability to do work.")

There is in physics an idea conveying notions of power, work, and force. It is enigmatic and filled with complexity. Even now not all its secrets are known. It is called energy and is the subject of this chapter.

Magic (Before Energy)

There is in the world a persistent belief in a "force" able to cast spells, divine, summon, and do "magical acts". Called magic, it is a common belief in primitive societies and in the occult. It was once thought that all natural features were a kind of magic that was controlled by God or the gods. It was thought magic manifested as Earth, Air, Fire, and Water. Also magic could be used to summon demons, cause rain to fall, to cure, to ward away "evil spirits", and so on. Today, magic is still believed in parts of the world and a hobbyist practice called magic (stage illusions and trickery) exists. Magic can be thought of as being a precursor to the concept of energy, however modern physics does not believe in magic actually existing.

Vis Viva (Energeia)

Along with magic, it was thought Nature was pervaded by an "energy" called "vis viva" (vees veevah). It dwelt in all material substances

FUNDAMENTAL CONCEPTS

and was believed in by the famous thinker Liebniz. It can be thought of as a precursor to the modern idea of energy.

Ether

There was a persistent belief through the ages that a "reality" or "universal substance" existed called "ether". It was thought to be a fundamental element, conveyed light waves, allowed for the influence of gravity, and had an unknown nature. Later on in the famed Michelson-Morley experiment, it was shown not to exist. However for centuries it was thought to be the source of energy.

Energy Fluid

Energy's nature has been mysterious for ages. Various thinkers (notably Benjamin Franklin) believed energy was a type of fluid (or liquid like water). They thought energy could flow (like a current), gather in deposits, and move about like other fluids. Later investigation showed energy was not a fluid and a nature different from a liquid.

The Nature of Energy

Energy is a vital idea to today's world. Complaints rage that there is not enough of it and that the world is running out of it. Qualities of its nature are the following:

- It exists –It is not magic, divine power, or any such thing –It is intimately involved with such physics ideas as work, force, power, kinetic energy, and so on –It is not intelligence nor a living thing –It exists throughout the universe –It can change from form to form and is conserved in interactions –Varieties of energy are X rays, gamma rays, electricity, magnetism, gravity, and so on –Energy factors in displays like explosions and lightning –It does not generate space nor time –Mass is a form of energy –It comes in five basic forces –It had an origin in the Big Bang –It obeys physical laws and theories –It is not mass directly –It travels at the speed of light –The total energy of the universe is constant –It can manifest because of motion as kinetic and potential energy –Energy is used in heating homes and powering modern technology –Energy is a phenomena – Energy can be controlled by Man for purposeful uses –Energy is measured in units called by names like the joule, electron volt, and horsepower

Types of Energy

Energy is known to come in many forms. These forms can change from one to another and always in a way where no more energy is created nor destroyed. This is referred to as the Law of the Conservation of Energy. Well known kinds of energy are the following:
- Electricity –magnetism –gravity waves –gravity –electromagnetism –nuclear energy –chemical energy –kinetic or mechanical energy –potential energy –elastic energy –X rays –gamma rays –light –heat –ultraviolet rays –infrared rays –radio waves – terahertz radiation –microwaves –long radiation –cosmic rays –TV waves -sound energy –vibrational energy –tension energy –atomic energy –field energy –internal energy

Please read about the various kinds of energy.

Concepts of Energy

Energy is today a vital and fundamental notion in physics. Its study is intense and has many basic ideas like the following:
- tension –work –power –current –resistance –efficiency –weight –mass –output –horsepower –erg –joule –candela –light –heat –equilibrium –luminosity –speed –speedometer –motion –field –force –gravity –potential –action –reaction -mechanical energy -tension -input -output -kinetic energy -potential energy

Please study the various ideas of energy to better understand energy.

Work

Work is an energy idea of being able to accomplish a task, move a mass object through a distance, and so on. Work as an idea now refers

FUNDAMENTAL CONCEPTS

to activities like writing, building, moving, pounding, washing, cooking, and doing tasks for pay or activity.

Potential Energy

It is known there is a version of energy dealing with potential or "before happening". It discusses the nature of energy before energy is actually used. Many energies like electricity, magnetism, and gravity have a potential energy counterpart. Tension in a string is an example of potential energy as well as the tension found in a compressed spring. When the string is snapped or spring is released, the potential energy is said to be released in an actual display of real energy.

Field

Its known that mass and charge are able to generate energy presences in space called "fields". Examples of fields are the electric, magnetic, gravitational, etc. Fields are known to generate "forces" which are pushes and pulls able to cause motion. Fields at times are considered a state of mass and are discussed in a subject called field theory.

Views of Energy

Energy is known to have had many views or visions as to what it is through the centuries. Presented here is a discussion on some of those visions.

- Superstitious energy. Energy was thought to be a magic of God or the gods.
- God's energy. It was thought God created energy or influences energy.
- Aristotle's energy. Energy was a consequence of the fundamental substances of Earth, Air, Fire, and Water.
- Earth energy. Energy was thought born from the Earth somehow.
- Ether energy. Energy was thought to be born of the "ether" or liquid reality.

- Newton's energy. Energy was the ability to do work, a force, a means to cause motion.
- Mach's energy. Ernst Mach, a teacher of Einstein thought energy was produced by the cosmic masses (or stars).
- Einstein's energy. Energy was a union with mass as "mass-energy", the ability to do work, a result of field and force, and so on.
- Mind energy. Energy is thought to be a creation of Mind.
- Psychic energy. Its thought "ghosts" can create or influence energy in some way.
- Particle energy. Energy is thought to have a particle nature and its particle is called the "ergon" and sometimes the "quantum".
- Science fiction energy. It is thought energy is born of a subatomic entity called the "string", a brane, a quantum fluctuation, or collision of universes.

Particle vs. Wave

For ages, thinkers have thought energy may come in particles or waves. Thinkers like Newton maintained energy (like light) was a particle and thinkers like Young maintained energy (like light) was a wave. Experiments were conducted that "proved" both sides were "right" at times. However in the 20th century, later thinkers like Planck and quantum theorists explored energy and found that energy had at times both natures. This caused a "crisis" that resulted in the discovery of the quantum and quantum mechanics. This realization that energy has both natures shook physics in revolutionary ways and continues on even now.

Quantum

Max Planck was a physicist about the time of Einstein. He took to studying the idea of the "blackbody" or an object colored black that does not emit any light, heat, or any other type of energy. He realized that energy comes in discrete lumps he called "quanta" (or packets). This discovery that energy had a particle-like nature shook physics and lead to the discovery of quantum mechanics.

FUNDAMENTAL CONCEPTS

The Changes of Energy

Energy is known to come in many varieties like light and heat. It is known that energies can change form or change from one type to another. An example is the photoelectric effect. This is the change of light into electricity and is done through the action of a device called a solar cell. There are many kinds of device in physics able to change one type of energy into another.

Models of Energy

Energy has been compared to such things to think about its nature like:
- fire –a wave –light display –mass –a piece of paper –river –line –string –rubber sheet

Law of the Conservation of Energy

There is famous law in physics called this. It briefly states that energy cannot be created nor destroyed, but only conserved from interaction to interaction. This means energy cannot be created nor destroyed, but remains in the same amount in whatever physical interaction (or event) it is involved with.

Philosophy of Energy

This is a branch of philosophy discussing the nature of energy. It discusses what energy is, how its produced, and how it relates to the nature of the universe.

Units of Energy

Its known energy is measured to according to various units. Some versions are the joule, kilocalorie, calorie, erg, volt, ampere, electron volt, horsepower, and so on.

Uses of Energy

Energy today is used for all kinds of activity. It is used to shine lights, power automobiles, run engines, run nuclear power plants, automate factories, produce electricity, run a marathon, and so much more. Famous devices of energy are the light bulb, electric power supply, atomic bomb, candle, flashlight, razor, blowtorch, flood lamp, lighthouse, and so much more. Energy is vital to all kinds of technology and new inventions using it are constantly being devised.

Energy Production

Energy is vital to modern technology, so much so that there is a worldwide movement to find ways to generate it. Topics of discussion on generating energy for useful power are the following:

- Geothermal energy (Earth's heat) –hydroelectric dams (rivers) –solar power (The Sun) –star power –tide power –current power –coal –liquid coal –wind (blowing air) –wave power –pedal power –atomic power –animal power –uranium –plutonium battery –breeder reactor –ionosphere power –Earth's magnetic field –fusion (join hydrogen nuclei together) –methane power (a gas) –tokamak (Russian fusion reactor) –biomass (decaying garbage) –whale oil –volcano –oil shale –hydrogen power –Dobereiner's lamp –blue energy –radioisotope thermoelectric generator (a generator that uses radioactive materials to cause heat and to cause power) –laddermill wind power –lightning –storm power –fuel cells –flow battery –thorium –novel batteries –nuclear waste power –thermal depolymerization (breakdown of chemicals by heat) –zero point energy (energy near absolute zero) –luminescence (anything giving off light) –sonoluminescence (light by sound)

Please read about these kinds of energy production in more definitive books.

Energy Crisis

Around the world as human populations grow, a problem is arising that the world is running out of energy. What this means is that too many people are using fossil fuels (oil, gas, natural gas, kerosene) and other energy sources in amounts more than what can be sustained.

FUNDAMENTAL CONCEPTS

This demand results in higher gas prices, higher prices, and inflation. Its thought the world will reach a state where oil will run out, fuel prices will rise dramatically crippling economies, and nations will be made desperate for energy. Today, many thinkers are hard at work exploring ways to fight the energy crisis. They are exploring ideas like fusion, biomass, geothermal, solar, wind, and nuclear energy. It is unknown if they can fight or stop the energy crisis in time, but efforts continue.

Fossil Fuels

Its known that in very ancient times, plant life covered the Earth. This plant life became "fossilized" and in time turned into substances called oil, coal, and so on. Later on in today's age, people took mining these "fossils" and using them for fuel hence the name "fossil fuel". In time, this lead to the creation of such substances as gasoline, kerosene, natural gas, shale oil, coal, and so on. People used them in great amounts and its thought their supplies would last for ages. Today, its known the world is running out of these substances and demand for them has grown immensely. This problem today causes the world energy crisis, the issue of global warming, and so on.

Combustion

Combustion is a technical term that refers to "burning", incineration, and so on. Its main phenomenon is called fire. Fire has been known for ages from primitive man to today's era. It has been used in dramas like cooking, heating, arson, fighting wars, acts of terrorism, acts of crime, acts of environmentalism, and so on. It is caused when chemicals extremely react leading to energy displays called "exothermic" (releasing heat). Fire has been called one of history's greatest inventions and it is a vital tool in modern society.

Geothermal

Various inventors have realized that places like geysers, volcanoes, and hot springs are natural sources of "hot water". They have built devices that harness this "geothermal" (heat of the Earth) energy to generate electricity.

Hydroelectric

Various inventors have built dams on rivers. They use the river's flow to generate electricity with turbines. Famous dams are the Hoover, Grand Coulee, and others.

"To ancient man, energy and magic were one and the same. Whoever was lord of one was lord of the other. The modern physicist is not a sorcerer, but through his mastery of energy seems like one."

-physics class discussion

Mystery Energies

In physics, many kinds of energy have been discovered, however there are continuing controversies that perhaps other kinds exist. Compiled here is a list of such controversies. The reader is encouraged to explore their issues more closely outside of this book. Examples are:

- Dark energy (unknown energy of the universe) –zero point energy –prana (India) –mana (Polynesian) –chi (China) –phantom energy –antigravity energy –vril –orgone (Wilhelm Reich) –odyle (Baron Carl von Reichenbach) –pyramid power (energy associated with pyramid shapes) –psychic energy (psychics) –cosmic energy (free energy from the universe) –Earth energy (free energy from the Earth) –Unruh radiation (William Unruh) –Hawking radiation (energy from self destructing black holes) –inflation energy (energy from the inflation expansion of the universe if real) –vital energy (the life force) –free energy –N rays (Blondlot) –aura energy (Kirlian) –spacetime energy –masstime energy –spacemass energy –anti-energy –negative energy –magic energy –Big Bang energy –superenergy –supernatural energy (energy from black or white magic) –kundalini (an energy discussed in Hindu belief).

FUNDAMENTAL CONCEPTS

Vital Force (Vital Energy-Vitalism)

There is a controversy in physics about the existence of an unknown energy called the "vital force". It is also called by such names as élan vital, bioenergy, chi, vital energy, kundalini, etc. A discussion in philosophy on it is called vitalism. It is thought living things have an energy around them or flowing through them that is "needed" to sustain life. Its been believed in for ages and in China it is a central teaching in such subjects like qigong and martial arts. Today, it has not been verified that it exists and is not currently believed in, but discussion and belief in vital energy continues.

Mana

In Polynesia there is a persistent belief in an "energy" called mana. It is thought to be a version of magic that was usable by Polynesian chiefs and holy men. With it, they could levitate large stones, move large objects, be able to persuade, or had "significance" of some sort. Belief in mana is persistent, but today its thought not to exist as magic is not believed to exist. Today, belief in mana is considered a superstition in Polynesia, a relic of ancient times.

Zero Point Energy

Its thought that about Absolute Zero, energy can be "extracted" (or taken) from empty space and this is called zero point energy (ZPE). Its thought ZPE could be a new source of usable energy, however it requires temperatures to be so cold to get it. For now, its thought it can never be used and it will remain only a possibility in physics.

Free Energy

Its thought in physics that Nature contains mysterious and unknown kinds of energy. These energies could then be used to power machines or could be used "freely" from Nature (hence "free energy"). Its thought "mystery energies" called Earth energy, cosmic energy, crystal energy, and so on exist and can be "tapped". So far there is no proof that they exist and hence they cannot be used. Many "inventors" have made outlandish claims that they have made free energy devices, but they cannot prove their claims. Today, the issue of free energy continues as a subject of pseudoscience, but also as an issue tantalizing physicists that perhaps new sources of energy exist unknown to Man.

Pyramid Power

Various thinkers have believed that the pyramid shape has energy properties unknown to science. They maintain pyramids can sharpen razors, preserve food, influence sleep and dreams, and so on. Called pyramid power, they maintain pyramids have an energy unknown to science. Its known that ancient Egyptians built many pyramids as well as other cultures. Its thought the Egyptians knew secrets about pyramids and used them for energy purposes unknown to science. While nothing has been proved to exist, pyramid power continues on as an intriguing issue. It is considered pseudoscience, but various thinkers maintain something is going on.

Negative Energy

Its thought energy has a counterpart that can be called "negative energy". Negative energy if real could cause antigravity, negative work, negative electricity, negative light and heat, and so on. While its unknown if it exists, it is a parallel idea to notions like negative mass, negative time, and negative space.

Energy has always been one of physics' most interesting issues. Probing the secrets of energy has gone on for ages and even now secrets still exist. No one knows what future thinkers will know about energy, but it is sure to be interesting.

In the following section, we will examine a concept often called the "twin brother" of energy. It had its origins in discussions on Earth and air and has become one of physics' most vital ideas. Let us now examine the nature of mass.

FUNDAMENTAL CONCEPTS

Matter

The Concept of Mass (or Matter, Koinomatter)

In physics, there is another fundamental concept and it is called mass or matter. Matter has been an enigma for centuries in that no one knew all its secrets. It is known to be a form of energy, yet it is distinct from energy in many ways. In this chapter, we will explore the nature of mass.

"Mass or Matter is Nature's most mysterious idea. Within it lurk secrets Man has never known and more is sure to be found."

-Voigt

Before Mass

The idea of mass had its origin in the Greek philosophy discussion called the four fundamental substances. The ancient Greeks believed in four substances called Earth, Air, Fire, and Water. Each substance was thought irreducible and hence "fundamental" in some way. In time, these beliefs would be overthrown as Water is composed of two gases called hydrogen and oxygen and the other substances would be found to have "smaller" more fundamental qualities. Mass would gain its modern conception with Isaac Newton when he devises the laws of motion.

Substance (Fullness, the opposite of emptiness)

There is in philosophy an idea called "substance". Substance refers to the material, fullness, hardness, solidity, reality, structure, and the opposite of emptiness. It is one of this subject's most classic ideas. Its thought the idea of mass had its origin in the idea of substance and this can be explored in works of philosophy.

The Nature of Mass

Mass is a complicated idea. It is a well-known idea used in physics,

THE GRAND SCIENCE OF PHYSICS

chemistry, and so many sciences. It has a vast nature and its qualities are the following:

- It exists –It is a form of energy –It has a "counterpart" in antimatter –When matter and antimatter meet, they self destruct in a display of energy –It has a composite nature in that it is made of atoms –Material bodies like stars and rocks are made of mass –Mass factors in conservation laws –Mass comes in a variety of states like liquid, solid, and gas –Mass is not space, time, nor spacetime –Mass is not a magic –Mass does not generate space, time, or spacetime –Mass is an idea of substance –Mass had a beginning –Moving masses are said to be in motion –Mass cannot travel faster than the speed of light –Mass factors in force interactions like gravity –Mass can have a weight –Mass objects fall at the same rate in a gravity field –Mass can be measured by means of scales –Mass is not a force, but is involved in the nature of force –Mass is not a living thing –Mass is a phenomena –Mass obeys laws of motion, physics laws, and physics theories –Mass is a fundamental idea –Mass behaves strangely near black holes –Mass can be controlled by Man for purposeful uses –Mass can "disturb" spacetime to cause gravitational fields –Objects of mass are the electron, proton, and neutron –Mass has many mysteries still –Elements are atoms of mass –Mass particles can have a quality called "charge" by which electricity can flow –Mass can have a particle or wave nature that is it can come in bits like grains of sand or undulations like water waves –Interactions of mass follow the Law of Conservation of Mass (it refers to a claim that mass cannot be either created nor destroyed, but is conserved from interaction to interaction).

Concepts of Mass

Mass is one of physics' most heavily studied ideas. It is known for many interesting ideas and a collected list of them are:

- weight –inertia –specific gravity –kinetic energy –acceleration –moment of inertia –center of mass –center of gravity

FUNDAMENTAL CONCEPTS

–work –impulse –torque –centrifugal force –inertial mass –gravitational mass –charge –substance –pressure –temperature –fullness -relativistic mass -tardyon -tachyon -ion

Please study the many kinds of concept of mass.

Momentum

Momentum is the idea of a mass in motion able to stay in motion. It is said to be conserved and has its own law of physics. It is studied in discussions on motion.

Fundamental Building Block of Matter

There is in physics a controversy as to whether there is a "smallest" bit or particle of mass. This has lead thinkers to conceive the idea of the atomos or "smallest bit of mass". Searches for the atomos have lead to the discovery of the atom. At one time, the atom was thought to be the smallest bit of mass, however it is known that the atom has yet smaller parts. Examples are the nucleus, proton, neutron, electron, and quark. Today, there are searches to find ever smaller particles, however today it is not generally agreed that there is a smallest particle. A speculation goes that perhaps quarks (particles that make up protons and neutrons) are made of yet smaller particles. These particles are so small in fact that they are the smallest things of all. Physicists like naming particles by attaching the suffix –on to a word, hence the idea of the smallest particle of all is dubbed the "smalleston". The smalleston is today only a speculation and is not known to exist.

Views of Mass

Mass has gone through several views or visions through the ages, presented here are some discussions on those views.

- Superstitious mass. Primitive men thought mass was a magic of God or the gods.
- God matter. Mass was thought to be a creation from God's magic.

- Aristotle's mass. Mass was thought to be a consequence of the fundamental substances of Earth, Air, Fire, and Water.
- Substance mass. It was thought mass may be a consequence of Wood, Metal, Spirit, Ether, or Quintessence, other kinds of "fundamental substance".
- Atomos mass. It was thought mass was made of point-like particles called atomos or atoms. These atoms are however not "modern atoms" as they are indivisible and thus the smallest particles of mass ever.
- Indian mass. In India, thinkers from Jainism and Hindu philosophy created their own versions of what an atom is and hence what mass is.
- Tao mass. Mass was thought to be a creation of Tao or the impersonal reality of Nature in ancient China. Yin-yang mass refers to mass coming pairs of some kind.
- Ether mass. Mass was thought to be born of the "ether", a liquid reality.
- Newton's mass. Mass was its own fundamental substance able to generate forces and separate from space, time, and energy.
- Mach's mass. Ernst Mach, a teacher of Einstein thought mass mysterious and somehow born of cosmic masses.
- Einstein's mass. Einstein thought mass was a state of energy by the equation $E=mc^2$.
- Mind mass. This is a controversial notion that mass is a creation of Mind.
- Science fiction mass. Mass has been speculated to have counterparts called dark matter, shadow matter, mirror matter, and so on. Its thought kinds of matter and particles exist like X, Y bosons, preons, smallestons, the Higgs boson, and so on. Its thought mass may be born of the "string", brane collisions, or collisions of universes.

"Mass has within it energy so powerful it could power a reactor or be used in a bomb. It took centuries to know mass was a source of energy. Einstein had to come along and say mass was a form of energy."

-class discussion comment

Antimatter

Various physicists named Dirac and Anderson had a suspicion that perhaps ordinary matter had a counterpart. This counterpart when it "interacts" with ordinary matter would cause the mutual self-destruction of both in a display of energy. This counterpart was named "antimatter" or the "opposite of ordinary matter". Since then, physicists have found that antimatter exists and particles of it are called the positron, antiproton, and antineutron. Its known from particle accelerators, that particles can be made from energy in pairs of both matter and antimatter. Today, antimatter is a curiosity in physics as physicists have constructed atoms of antimatter (called antihydrogen). It is for the most part a subject of science fiction thinkers in the conceiving of star drives and spaceships.

States of Mass (or Matter)

Mass is known to come in many varieties. These varieties differ in terms of temperature, pressure, or condition. Compiled here is a list of known states of matter and its believed even more states will be discovered.

- Solid –liquid –gas –plasma (ionized gas)–supercritical fluid –quasi-solid –liquid crystal –QCD matter –quark liquid –strangelets –singularity (a mass of infinite density found in a black hole) –antimatter (a counterpart to matter like the positron, antiproton, and antineutron) –amorphous solid (substances like glass) –neutron star matter –quark gluon plasma –photonic matter –neutrino matter –Bose/Einstein condensate (also called the "superatom") –fermionic condensate –superfluid (a super-cold liquid that flows without resistance) –supersolid (a super-cold state of mass) –nucleonic matter (particles like the neutron, quark, and proton) –white dwarf matter (matter found in a white dwarf star) –electronic matter –field (an energy presence in space radiating from a mass, but also considered a state of matter) –weakly symmetric matter –strongly symmetric matter –degenerate matter (matter found in compact stars) –quantum spin Hall state matter – strange matter –string net liquid –metallic hydrogen (hydrogen under intense pressure that it exhibits metallic qualities) –neutronium –crystalline solid –plastic crystal

Water is a good example of what a state of mass is like. Ordinary water is a fluid that comes from a faucet or is used to wash dishes, this is referred to as a liquid. If you boil water, it can turn into a gas called water vapor. If you freeze the liquid water, it becomes a solid called ice. Other states of matter are similar to this depending on their temperature (how hot they are) and pressure. States of matter tend to be technical discussions and are for the most part the domain of research for the professional physicist.

Please read about the various kinds of state of matter. Its thought other kinds of state will be discovered in the future.

Manifestations of Mass

Mass is known to make up much of the natural world. It results in so many kinds of material and even now more kinds of material are being found. Mass seems to be everywhere and in everything. Various "common manifestations" of mass are:

- clouds –water –air –rock –minerals –gems –plastic –flesh –blood –leaves –wood –plastic –metals –acids –bases –stars –planets –asteroids –comets –dust –sand –fossils –lava –wool –fur –cloth –plexiglass –glass –rubber –salt –gold –silver –mercury –ice –superfluids –supersolids –cardboard –paper –hair –bone –shells –concrete –smoke –fire -particles -gems -singularity -radon -helium

Changes of State

Mass is known to be able to change state or phase. Topics of

FUNDAMENTAL CONCEPTS

discussion here are freezing, thawing, melting, boiling, subliming, vaporizing, ionizing, and so on.

The Measure of Mass

Mass is a quantity in physics that is measured by use of scales or balances. It comes in units like grams, kilograms, slugs, and so on. Determining the mass of an object is a main activity of physics.

Weight

Weight is another vital concept in physics. It is defined as the product of mass with the acceleration due to gravity (W=mg). There is a confusion in physics that the measure of weight is also the measure of mass. These two ideas are very different and should not be identified with each other.

Density

This refers to the idea of "stuffing" matter into a volume. An example is a balloon. An empty balloon has very little air in it and thus it has a "low density" of mass in it. When it is filled with air, it "inflates" and is thus said to have a "high density" of mass in it. Density is a basic idea in physics and chemistry referring to "stuffing" mass.

Gravitational Mass

Its thought mass somehow generates gravity. This has lead thinkers to invent two kinds of mass able to generate gravity, dubbed active and passive gravitational mass.

Mysteries of Mass

Mass is a subject filled with mystery and controversy. There are in physics many issues relating to mass. Examples are that mass has more versions (or states) than are known or that It has mysterious properties unknown to science. Issues for research and discussion are:
- Dark matter (unknown kinds of matter in the universe) –the missing mass of the universe –tachyons (particles that go faster

than light) –superbradyons (types of tachyons) –the Higgs boson (Peter Higgs) –undiscovered particles like preons, X bosons, and other entities –shadow matter (another kind of counterpart to matter) –mirror matter –negative mass –imaginary mass –supernatural mass (mass controllable by magic) -pre-cosmic matter (mass of another universe) –neutralino –axion -supersymmetric matter –chargino –Goldstone boson –goldstino –graviton –gravitino –photino –dilaton –magnetic monopole –unifon (particle of the superforce) –smalleston (smallest particle of mass) –pentaquark –Oh My God particle –dilatino –axino –Wino –Zino -crypton

Inertia ("Resistance to change")

There is a concept in physics deeply connected to mass, it is called inertia. It refers to the notion that masses have a "built-in reluctance" to change shape, nature, or direction of motion. It is in itself not a force nor a measure of mass, it is overall an issue derived from discussions of mass. The reader is encouraged to explore its issue more closely.

Subatomic Particles

In the early 20th century, physicists like Ernest Rutherford, JJ Thomas, Chadwick, Niels Bohr, and others probed the atom for smaller entities. This lead to the discovery of such particles like the proton, neutron, electron, the atomic nucleus (or center mass of an atom), and so on. In time, many more kinds of particle were found. This began a drama in physics whereby scientists build large machines called "atom smashers" to look for ever smaller particles. Today physicists know about all kinds of small particle called by such names as neutrino, photon, gluon, quark, etc. Subatomic particles are known to come in "groups" variously labeled as boson, fermion, hyperon, meson, hadron, lepton, baryon, etc. To this day, many more kinds of particle are being discovered and many Nobel prizes have been given over their finding.

FUNDAMENTAL CONCEPTS

Higgs Boson

There is in advanced physics a strong belief that a particle (still undiscovered) exists able to cause mass to exist. Called the Higgs boson (or bosonic particle of Peter Higgs), it is being sought after in particle accelerators about the world. Its thought it must exist to explain features of advanced physics and its discovery is an eagerly awaited event in physics today.

Philosophy of Mass

There is a branch of philosophy dealing with the discussion of the nature of matter. It discusses issues like substance, atomos, the fundamental building block of matter, and so on. Please explore this subject in other books.

Crystals

Sometimes mass can be made to grow or collect in "solid arrangements" called crystals. Crystals are mass collections known for order, symmetry, regularity, hardness, and precision. Crystals come in kinds like ice, quartz, diamond, emerald, ruby, sapphire, gems, and other substances. Crystals can be grown from atoms. The study of crystals and their structure is called crystallography. Crystal structures are called lattices.

Fluid Flow

Fluid flow has been a subject of intense study for ages. It began with watching rivers flow and in time grew into an intense subject. It now discusses such issues as designing dams, the flow of honey, the flow of gases, the flow of molasses, the flow of most any kind of fluid in Nature (like water, liquid helium, mercury, acids, bases), viscosity, and so on.

Gas

A gas is a state of matter known for blowing, flowing, being felt as wind, assuming the shape of its container, and being different from a liquid or solid. Gases come in kinds like air, noble gases, oxygen, hydrogen, helium, nitrogen, and so on. Gases make up the atmosphere and are intensely studied in chemistry.

Atmosphere

Around the Earth, there exists a layer of gas that extends into space. Called the atmosphere, it is a layer of air consisting of nitrogen, oxygen, carbon dioxide, argon, neon, and other gases. It comes in layers called the troposphere, stratosphere, mesosphere, ionosphere, thermosphere, and exosphere. It contains a layer of oxygen called the ozone layer that protects Earth from ultraviolet radiation. It is intensely studied in science as its known for wind, weather, clouds, storms, and so on.

Solid

A solid is a state of matter known for being hard, having a structure or crystal quality, being immovable, not flowing, and being different from liquids and gases. Types of solids are crystals, dust, rock, minerals, glass, plastic, fabric, and so on. Solids are studied in chemistry and by passing radiations through them to determine structure.

Plasma

A plasma is a state of matter that is like a gas, but is known to be extremely hot (they are called ionized gases). Types of plasma are lightning, fire, "glowing gas", aurora, and so on. Plasmas are intensely studied in plasma dynamics and chemistry.

Superfluid

A superfluid is a state of mass that occurs near absolute zero. Helium, a gas when cooled to near this temperature changes into a superfluid. A superfluid is not like an ordinary liquid, it is known to be very cold, to flow uphill, and has many other qualities. Its known only a few gases are able to be changed into superfluids.

FUNDAMENTAL CONCEPTS

Bose-Einstein Condensate

In Einstein's time, a thinker named Bose (Bo zee) conceived the idea of a state of matter unknown to physics. Einstein improved on his work and they derived the notion of the "Bose-Einstein Condensate". The BEC is a state of matter where it seems all a particle's nature is concentrated at a point also called a "superatom". Many years later, thinkers like Ketterle and others made a real example of a BEC proving that Bose and Einstein's idea was in fact reality. This lead to the awarding of a Nobel Prize and a research fad to try to invent other kinds of "condensate".

Singularity

When a star is crushed by its own gravity, it can form an object called a black hole. Inside the black hole, mass is crushed into an object of infinite pressure and density called the "singularity" (or one object). The singularity can then be looked upon as a state of mass unlike any other. It is crushed so tightly, it is in effect a point (or geometric point). Its thought the universe began as a singularity and that inside every black hole there exists a singularity. Its thought a singularity cannot exist outside of a black hole as it takes a black hole's intense gravity to create one (a naked singularity).

Tachyon

There is a speculation in physics that perhaps "objects of mass" exist able to travel faster than light. Called the tachyon, this particle is thought not possible to exist, but others have sought to detect it anyway. If the tachyon exists, it is believed it can travel at near instantaneous speeds anywhere in the universe, that it always travels faster than light, that it takes an infinite amount of energy to slow it down to the speed of light, and so on. Its existence has been a favorite topic for science fiction thinkers as its used in fantastic kinds of starship. For the time being, its existence is a kind of fantasy in physics. Its thought impossible to exist, but physicists hope that maybe it does exist.

Negative Mass

Its thought "positive mass" (ordinary mass) has a counterpart called "negative mass". Just as positive mass can cause gravity, its thought negative mass could cause antigravity. For the time being, its not thought to exist, but its hoped that it may one day be really. Ideas like antimatter, shadow matter, mirror matter, and dark matter are not thought to be negative mass.

String

In advanced physics there is a controversy over the possible existence of a subatomic entity called the "string". It is a "line" of spacetime able to bend, twist, vibrate, and act in such ways as to create mass, forces, and energies. Today it remains undiscovered, but a research fad has grown up called superstring theory discussing it. For now, it remains as a tantalizing possibility in that it could cause mass to exist.

Higher Dimensions

Its thought spacetime may have more dimensions than three and one of time. Its thought the universe exists within a "reality" of higher dimensional space. The actions of these higher dimensions is speculated to cause mass to exist.

Releasing Energy

Its known that mass contains immense amounts of energy. Scientists have tried for ages to find ways to get mass to release energy. Examples are radioactive decay, a nuclear chain reaction, a nuclear explosion, energy generation, fusion, etc.

Creation of Mass

Mass is said to be be conserved. This means it can be neither created nor destroyed. However, there is a continuing mystery of how the universe acquired mass and somehow energy became mass after the Big Bang. No one knows how mass acquired its nature. Today, its still very mysterious as more is being learned about it.

FUNDAMENTAL CONCEPTS

Mass overall is a subject that has fascinated many thinkers. Probing the secrets of matter has been one of physics' longest activities of research. Today, it hides secrets still and it still continues to fascinate physicists even now.

Mass and energy have a counterpart like spacetime, this idea is known as massergy. Massergy is the union of mass and energy as one. Its nature is similar in ways to both mass and energy and hence its discussion involves a blend of both. In the next section, we will briefly review this important, but often obscure idea in physics.

Massergy (Mass-Energy)

The Concept of Mass-Energy (or Massergy)

This concept is similar in many ways to Mass and Energy. It factors in discussions on the Law of the Conservation of Mass-Energy. For the most part, it is fairly obscure in physics and its discussion is similar to mass and energy.

Law of the Conservation of Mass-Energy

There is a law in physics that says that neither mass nor energy can be created nor destroyed, but only conserved from interaction to interaction. What this means is that mass nor energy cannot be created nor destroyed in a physical interaction. The amount of both you have at the beginning is the same at the end. This law has been extended to mass-energy when the two concepts were unified as one.

> "Einstein showed that mass was a form of energy and this is demonstrated by an atomic explosion."
>
> -class discussion

In the next section, we will review another basic idea of physics. It is most often found in philosophy, but it is critical to physics as well.

It involves the idea of thinking about physics and this subject is collectively known as 'mind'.

Mind

"Without a mind able to know, a physicist then will be unable to understand physics."

-class discussion

The Concept of Mind

Mind is an idea that represents "thoughts". In physics, students are taught all kinds of idea like mass, energy, space, time, inertia, force, weight, work, and so on. Each idea is a concept in Mind. Without these concepts, physics could not be done and thus would not exist. For ages, Man has had a "conception" of what the universe was and this has changed over the centuries. Mind is found throughout physics and science wherever concepts are taught and learned.

Issues of Mind

Mind is a fascinating issue in science. It refers to the ability to comprehend or an entity of self that comprehends. Various famous issues in mind discussion are:

- Imagination is the ability to conceive in original ways or in fantasy.
- Concepts are ideas or constructions of mind. Examples are space, time, mass, energy, order, harmony, shape, power, current, duration, etc.
- Intuition refers to somehow knowing something or being able to "know" magically.
- Reason refers to using thought to order or to understand order.
- Comprehension refers to understanding or somehow knowing something.

FUNDAMENTAL CONCEPTS

- Sense refers to abilities of body to feel or know about things like sight, hearing, taste, touch, and smell.
- Sensation refers to experiences like color, saltiness, wetness, hotness, coldness, hardness, and other "sensations" of senses.
- Explanation refers to a framework to understand something, a reason for.
- Description refers to discussing details about a thing.
- Insight refers to knowing the inner workings of something.
- Words are statements that represent a thing in Nature. Examples are space refers to emptiness and vacuum, time refers to the flow from future to past, etc.
- Language refers to using words to form thoughts and sentences to communicate.
- Meaning refers to the understanding of words and sentences.

Please explore the science of psychology for other words involved with mind.

Thought and Concept

These are words that refer to "thinking" and "ideas that are thoughts". In physics, concepts (or ideas in mind) are everywhere. Basic physics ideas are space, time, mass, energy, massergy, force, inertia, work, power, law, motion, current, gravity, and so on. Without concepts, physics as it is presently taught could not be done. Today, the nature of concepts are taught throughout physics and are vital to its discussions. There is a science fiction notion that perhaps thought is its own substance, that it can be harnessed to do acts of the paranormal, or that it is its own fundamental force. For the time being, its believed thought is not a force, but is a consequence of brain and electrical disturbances.

Consciousness

This is an idea that refers to "mind's thinking, awareness, identity, and personality as a unity". Consciousness has always been a mysterious subject in science. It is found in psychology, parapsychology, and in quantum physics discussions. It is thought consciousness arises from the brain and has its source in electric disturbances in the brain. For now, it's a controversial issue entangled with issues like the paranormal, universal mind, anima (spirit-soul personality), and other debatable topics.

Dictionary

A dictionary is a book of words of a language. It is also a book that contains many concepts expressed as a word. Each word in a dictionary is also an expression in mind.

Mind Over Matter

In studies of the paranormal and physics, there exists this issue where its thought "mind" has some "power" to control "matter". That is there is thought to exist some "interaction" or "force" that allows mind decisions to control material, worldly qualities. While the paranormal (especially the issue of psychokinesis) has been thought to be mind over matter, there is no proof that any paranormal ability exists and thus it is not believed in by physics. For now, mind over matter lies as an obscure, but controversial issue in physics. This is studied in parapsychology. The paranormal is also called the psychic or psi. Issues famous in the paranormal are the following:

- reincarnation -astral projection -out of body experience -prophecy -clairvoyance -intuition -conscience -near death experience -premonition -past lives experience -psychokinesis -empathy -bilocation -summoning -levitation -rapping -haunting -thoughtography -psychic archaeology

Universal Mind

There is in physics and philosophy an obscure idea called "universal mind". Universal mind refers to a "reality of mind" like a parallel universe that is somehow able to "think" and thus influence "physical reality" (the universe). Universal mind has also been called "cosmic consciousness", supreme intelligence, the Mind of God, and so on. So far there is no evidence that a universal mind exists and it is not believed to be real in physics. Please explore philosophy books on "mind".

FUNDAMENTAL CONCEPTS

Knowing

This is the idea and discussion on what it means to understand or "comprehend". It refers to mind's ability to think, conceive, create, define, and 'understand'. It is a bewildering issue, but nevertheless its idea is important in physics. Physicists maintain they "understand" ideas, theories, concepts, forces, and phenomena. They are in some sense "knowing" what physics is and what "physical things" are. Discussion on "knowing" is intense and is a topic in philosophy and physics.

Mind Philosophy

This is a sub-issue in philosophy concerned with thinking, idea, comprehension, and the nature of mind and thought. It has many thinkers, treatises, and related works. Please explore this subject outside of this book.

Psychology

Psychology is the science of mind, brain, behavior, and so on. It is known to study consciousness, thought, feeling, idea, sensation, and other issues. It is taught in many colleges and there are many works on it. Famous thinkers in psychology are Freud, Adler, Eriksen, philosophers, Babbage, modern thinkers, and professors.

Please study the many books on psychology.

Brain

Within the human body and many kinds of animal there exists an "organ" called the brain. It is composed of cells and is considered the center of control and mind in the body. It is thought that mind somehow originates with the brain, but for now this issue is filled with too much mystery and controversy. Please study the various discussions of brain in psychology books.

Language (Linguistics)

Various people communicate with each other through a system of words and sentences. These systems are called languages and the world

knows many kinds like:
- English –Nordic (Icelandic, Norwegian, Swedish) –German –Romance (Spanish, Latin, Portuguese, Provencal, Italian, French) –Greek –Turkic –Hungarian –Polish –Russian –Bulgarian –Hebrew –Arabic –Swahili –Xhosa –Slovak –Czech –Serbo/Croatian –Farsi –Dari –Pashto –Kannada –Punjabi –Bengali –Korean –Chinese –Japanese –Thai –Malayalam –Hindi –Hawaiian –Lakota –Navajo –Hopi –Havasupai –Oneida –Seneca –Athabascan -Marathi -Tibetan -Dzonghka -Malayalam

There are in reality many kinds of languages, sign languages, alphabets, grammar, words, signs, gestures, and qualities that communicate a message of some kind.

Mind overall is a contentious issue in physics with a lot of mystery. But it has importance in physics in that physics' basic ideas can be expressed as ideas that can only be understood by a mind.

> "A field is like gravity, its pulling is felt throughout space. A force is the pull you feel as you move in a field. Field and force are not the same. They are like a brother and sister, related but different, close but not identical."
>
> -class discussion

In the next section, we will explore the nature of two words used throughout physics. They factor in discussions on gravity, nuclear forces, EM, and so on. Even now, physicists are hard at work trying to unify all the forces as the result of just one fundamental force (or superforce). We now enter the discussion on two concepts known as field and force. A field is an energy presence in space caused by charge, a mass, or a particle-energy entity. A force is the result of a push or pull (or is it directly) that occurs when a mass is in the presence of a field. Even now all their secrets are not known and they continue to be fascinating issues in physics research.

Chapter 4
Physical Fields and Forces

"Space has a presence in it more than just the emptiness it is."
—class talk

Field And Force

Field and Force

There are two ideas in physics fundamental to the discussion of modern forces. They are called field and force. A field is an energy presence in space caused by some particle. For example, an electron has a quality called "negative charge", this creates an "electric field" in space. A force is a "push" or "pull" as felt by a particle of mass when it is in a field. A force can also be referred to as an interaction or the idea of fields meeting each other. For example, a proton has a quality called a "positive charge" and hence creates an electric field. If an electron enters a proton's electric field, it "feels" a "pull" to approach closer to the proton. This can be thought of as a force. Gravity is a force generated by mass to pull other masses near to it as demonstrated by a falling leaf falling towards the Earth. The Earth "exerts" a force on the leaf to pull the mass of the leaf closer to the Earth. In the next chapter, force and field factor in discussions on five entities called fundamental forces. Today, they are hot topics in physics at the forefront of physics research. The five forces are incidentally called gravity, electromagnetism, the weak nuclear force, the strong nuclear force, and the electroweak force. We will explore their discussion in the next section.

"There were thought to be many kinds of force in Nature each as fundamental as the next. However physicists found some were illusions and some were made of other forces. Nowadays, thinkers believe there is only one force in Nature and all the others are just illusions or derivates of the ultimate force, the supreme force of the universe itself."

-Voigt

Forces of Nature

It was once thought Nature was filled with all kinds of fundamental forces. Examples were friction (rubbing things together), celestial gravity (gravity of the stars), terrestrial gravity (gravity of the Earth), heat, light, electricity, magnetism, nuclear force, atomic force, chemical force, lightning force, motion force, time force, centrifugal force (spinning masses), centripetal force, Coriolis force, aurora force, tidal force, and so on. In time, scientists discovered many of these forces did not exist or were aspects of another more fundamental force. Eventually, all the forces of Nature were "reduced" to just four forces or now just three or five depending on how you look at it. For the time being, these are the forces that scientists have found.

In the next section, we will examine the nature of the forces known to physics and explore their intricacy.

Gravity

Gravity (Force) and Gravitation (Effect of Gravity)

There is in physics a force known for attracting masses to each other. It is responsible for falling leaves, orbits, free fall, and any type of falling really. Its nature has been mysterious for centuries occupying the attention of such thinkers as Aristotle, Newton, Einstein, and many others. Nobody really knew its cause or the secrets of its nature, however many "views" have been proposed on it. Examples of such beliefs are that gravity was a "downward seeking tendency of the fundamental substance of earth", a magic, a force of mass, the effect of the curvature of spacetime, the effect of a particle of gravity, the effect of a gravitational "ether" or "liquid reality", and so on. Gravity's riddle has

FIELD AND FORCE

gone through many theories and today it is thought to be the effect of mass "warping" spacetime to produce curvature (or bending of spacetime like bending a piece of paper). Today the best theory at explaining it is called general relativity, however there is a struggle in physics to find an even better theory of it. Attempts have been made to unite this theory invented by Albert Einstein with quantum mechanics, this has produced the drama called quantum gravity research. Its thought that gravity has a counterpart that is able to cause flight and this is called antigravity. It is thought however antigravity does not exist, but efforts to find it are called gravity research. Today, it is thought physics' understanding of gravity is incomplete and physicists are still struggling to understand this most mysterious of forces.

Textbook "Gravitation" by Wheeler, Thorne, and Misner

In advanced physics, there is a textbook written by the trio of Wheeler, Thorne, and Misner dealing with gravity called Gravitation. It is considered the premier authority on this force of Nature and should be studied for a deeper insight into this force. There exist many textbooks on gravity and the reader is encouraged to explore the various works.

Gravity Extras

Gravity is a force with an enormous literature to it with many topics. Explorers of gravity have attempted a lot of things like composing new theories of gravity, unifying it with other forces, trying to cause weightlessness, and so on. Gravity is known for a number called the gravitational constant and Newton's law of gravity. This is an equation that tells how to calculate the force of gravity. Various thinkers believe a graviton or particle of gravity exists, but this cannot be proven. Many attempts to make a particle theory of gravity have occurred as championed by the thinkers LeSage and Fatio, but this has come to nothing. The idea of a repelling gravity or "repulsity" has been proposed, but no such force has ever been found. A famous organization to explore in connection to antigravity is the Gravity Research Foundation. Gravity

THE GRAND SCIENCE OF PHYSICS

today is a profound idea in physics and topics to explore on this force are the following:
- gravity assist (using planet's gravity to give a push to space probes) –gravitational lens (using the gravity of galaxies to concentrate light) –gravity technology (pits, pipes) –gravitational astronomy (using gravity to bend light and study distant objects) –gravity waves –quantum gravity –black holes –gravitational collapse –triple axis –artificial gravity (making gravity happen) –weightlessness (an experience of no gravity) –antigravity (a force of pushing) –antigravity shields –gravity shields –gravity control –negative mass –gravitomagnetism –levitation (float flight) –electrogravity (controlling gravity by electricity)

Up and Down

These words reflect what gravity means to most people. Up refers to a direction away from the Earth and towards the sky. Down refers to a direction towards the ground and center of the Earth. In space, gravity does not exist and so up and down have no meaning.

Views of Gravity

Gravity has been a controversial topic in physics for ages. Its nature has mystified thinkers from Aristotle to Einstein. Presented here are discussions on how gravity has been viewed or envisioned through the ages.
- Superstitious gravity. The ancients did not understand the nature of gravity. They knew that it was an influence that pulled things down (like falling leaves). Various thinkers thought gravity was a magic of the gods, a magic of witchcraft, a spell of a divinity, a natural magic, a magic of Earth, or a power of God causing things to fall.
- God gravity. This is an ancient notion that gravity is caused by the power of God.
- Aristotle's gravity. Aristotle was a major thinker of ancient Greece. He thought gravity was a force that pulled things down to their "natural place". He also thought heavier objects fell

FIELD AND FORCE

faster than lighter ones (rocks falling faster than feathers). He believed gravity arose from a fundamental substance like Earth, Air, Fire, and Water.

- Vitruvius gravity. Vitruvius was a famous Roman thinker who maintained gravity did not depend on the amount of substance, but the nature of a substance. He could not explain how substance (later called mass) could cause gravity.
- Brahmagupta gravity. Brahmagupta was a famous ancient Indian thinker who maintained gravity originates with the Earth, but he could not explain why.
- Musa gravity. Musa was a famous ancient Arab thinker who maintained gravity was a force of attraction, but its origin was unknown.
- Al Khazini gravity. Al Khazini was a famous Arab thinker who thought gravity resulted from a pull from the center of the Earth. He could not explain how the center of the Earth could cause a pull.
- Galileo's gravity. Galileo challenged Aristotle's views on gravity and believed masses fell at the same rate. He did not know the nature of gravity either and thought it was a force of some kind.
- Newton's gravity. Newton extended on Galileo's notions of gravity. He reasoned that celestial gravity and Earth gravity were one and the same as universal gravity. He also thought gravity was a force and devised a famous force law ($F = GMm/r^2$).
- LeSage gravity. LeSage was a thinker near to Newton's time who believed gravity was a force caused by particles. He could not prove his idea and his notions are today an obscurity in physics. Similar views were also held by a thinker named Fatio.
- Ether gravity. Various thinkers thought gravity was caused by a "liquid reality" called ether, also used to explain light as well. No proof of this has ever appeared and its not currently believed in.

- Mach's gravity. Ernst Mach, a teacher of Einstein thought gravity originated from the cosmic masses (or a mysterious interaction between all the stars in the universe).
- Einstein's gravity. Einstein came along in 1915 and proposed the theory of general relativity. He united space and time as spacetime. He also said that gravity was the result of the "curvature of spacetime". Today, Einstein's views on gravity are physics' main belief, but there are hints that perhaps he will be replaced someday.
- Graviton gravity. Modern thinkers have speculated that perhaps gravity has a particle called the graviton. They invented this in order to make a quantum theory of gravity, a theory so far undiscovered in physics. Today, it is not known if the graviton exists, but searches for it continue. Other thinkers have taken to inventing other kinds of gravity particle going by names like gravitino, sgraviton, and sgravitino.
- Science fiction gravity. Modern thinkers in physics have taken to fantasizing other ways gravity might originate. Ideas range from "brane collisions" to collisions of other universes with each other. Today, its not known if such views are real, but speculation continues.
- Higher dimensional gravity. Modern thinkers have taken to speculating that gravity originates with dimensions of space unknown to physics.
- String gravity. This is a controversial notion that perhaps a subatomic entity called the "string" exists. By its waves, vibrations, twists, and other behaviors, then the various forces and particles of Nature are made. It is thought the string may create gravity.
- Mind gravity. This is a controversial notion that perhaps gravity is created by Mind.

General Relativity

In 1915, Einstein would propose a theory called general relativity. It would overthrow Newton's theory of gravity. He claimed gravity arose from curved spacetime caused by masses warping spacetime. Since then his theory has been tested by many and it has caused acclaim throughout physics. Today GR is physics' best theory of gravity, but efforts have been made to replace it. So far, GR continues on with predictions like frame dragging, gravity waves, and other discoveries. It is a testament to the genius of Einstein at the greatness of theory it is.

Arrow and Speed of Gravity

Its known gravity "pulls" from up to down and its influence travels at the speed of light. Why gravity pulls is unknown and since gravity is an energy, its influence travels only at one speed, the speed of light.

Acceleration Due to Gravity

Its known that objects in free fall feel a "tug" by gravity. Objects then are caused to move and fall down. Its said they are accelerated to fall down. There is a constant in physics called "g" or the acceleration due to gravity. It factors in calculations of weight. Its value on Earth is 9.81 m/s*s. Its value is known to be different on various mass objects. Escape velocity is the idea of the speed a mass object requires to escape the pull of gravity.

Graviton vs. Curved Spacetime Debate

There is a controversy in physics that gravity results from either a particle (called the graviton) or curved spacetime (Einstein). Thinkers trying to come up with the mythical theory of quantum gravity are trying to "quantize" gravity so it can be joined with general relativity. So far this effort has not achieved success, but beliefs in the graviton are "confident" in a physics research area called "quantum gravity".

Gravitational Constant (Force of gravity)

Isaac Newton invented a number called 'G', it is the symbol for the gravitational constant. It is a special number used in the gravity force law ($F=GMm/r^2$) to calculate the force of gravity between two masses. There is a modern speculation that perhaps G can change its value thus affecting the nature of the universe.

Newton's Gravity (Universal Gravitation)

In Newton's time, it was thought that the gravity of falling leaves (Earth or terrestrial gravity) and the gravity of the Sun (celestial gravity) were two distinct forces of gravity. Isaac Newton came to believe that both these "types" of gravity were in fact the same. He realized they were just different manifestations of a "universal gravity force" and from this he composed the famous law of gravity.

Gravity Waves

Einstein in his theory of general relativity predicted the existence of gravity waves. They are an altogether new form of energy previously unknown in Nature. Einstein's prediction of this energy lay neglected in physics for decades. Later on, the team of physicists Hulse and Taylor studied a binary pulsar (two neutron stars in orbit about each other). They discovered that this pulsar was emitting gravity waves and hence they discovered gravity waves. They were given a Nobel Prize over this and this issue has since opened the subject of gravity wave astronomy.

Gravitational Collapse

Its known in physics that a star's gravity is what holds a star together. It was thought by such thinkers as Michell, LaPlace, and others that if a star had enough mass then its gravity would cause it to collapse violently inward. It was thought that there was no force that could save a star from being crushed by its own gravity. It was thought perhaps stars could be so massive that they could crush themselves out of existence or into small, highly concentrated stars (compact stars). This has lead thinkers to invent notions like the neutron star, white dwarf star, and black hole to describe stars that are crushed by their own gravity. Today, this issue of gravitational collapse is a major topic in physics with an intense discussion.

Antigravity

There is a controversy that perhaps gravity has a "counterpart" force that "repels" or "pushes up" instead of pulling down. Various thinkers have thought this force exists, but searches for it have found nothing.

Today, its not believed to exist, but its hoped some version of it can be found. Antigravity has been called by many names through time and various names are:
- repulsity (repelling gravity) –antigravity –countergravity –levity –levitation –floating force

Quantum Gravity

Physicists today are trying to understand gravity of its deepest secrets, but presently cannot do so. Its known today's theory of gravity is general relativity by Einstein. Many physicists are trying to join GR with quantum mechanics to form a quantum theory of gravity. Supposedly if this theory exists, then a new understanding of gravity will be. So far this theory cannot be found, but efforts are underway to find it.

Unifying Gravity

Many thinkers like Einstein wanted to unify gravity with other forces. This has lead to research efforts like quantum gravity, "EMgravity" (unifying gravity with electromagnetism) or "electrogravity", strongravity (unifying gravity with the strong nuclear force), and so on. So far nothing has worked as no one has been able to unify gravity with any other force or combination of forces.

Particle Gravity

Its been thought for ages that gravity had a particle nature. This idea was begun by thinkers named LeSage and Fatio about Newton's time. Eversince, this notion has had bouts of research by many thinkers. Today, its thought gravity results from curved spacetime and is described by Einstein's theory of general relativity. Attempts have been made to revive particle gravity by suggesting the graviton and gravitino exist. Thinkers into quantum gravity are trying to construct a quantum theory of gravity and so far cannot do so. For now, particles of gravity remain undiscovered.

Gravity Shielding

Many thinkers have wanted to invent a "shield" to block against the influence of gravity. It was hoped this would lead to antigravity, but this presently cannot be achieved. Gravity has the effect of being able to pass through whatever mass barrier there may be.

Gravity Control (Artificial Gravity)

Many inventors have wanted to know how to control gravity. They have wanted to be able to make intense gravitational effects, reduce gravity's effects, or negate gravity's pull altogether (antigravity). So far they have not had much success, but efforts to do so continue. Its thought gravity could be controlled by using other energies, however no way is known of doing that however. For now, gravity remains an enigma as finding ways to make antigravity technology and other tools of it seems to go nowhere for now.

O'Neill Colony

Gerard K. O'Neill is a thinker famous for inventing a kind of space station. It is a wheel spinning about an axis, but none has ever been built. When it spins, a centrifugal force is created that can simulate gravity.

Gravity Propulsion

Various thinkers believed if gravity could be controlled to be a repulsive force then new kinds of technology could be created. This would lead to new kinds of propulsion (like a rocket) that could move a spaceship or move anything else. So far research into this has not gone far, but it is known the US government and various research organizations look into it.

Weightlessness

In orbit, Earth's gravity is so weak a state comparable to antigravity is created. Here astronauts aboard spaceships can "float" or "levitate" as gravity is "nullified" or made not to exist. It is thought that perhaps a futuristic space station called the O'Neill colony can be made. It is a giant ring (or torus) that can be made to spin like a centrifuge. This spinning object can in some sense make a centrifugal force effect which

FIELD AND FORCE

can simulate the effect of gravity. Its thought gravity's effect is similar to centrifugal force or acceleration.

Levitation

Levitation is the idea and effect of "floating" in air. It has long been the dream of ages of Man being able to fly like Superman, the birds, and bees. This has lead thinkers to try to discover antigravity, to find some way to simulate flight, and try to get into the air by balloon or some other means. In physics, various thinkers have found that magnetic fields can be used to cause "levitation" of objects like trains, this is referred to as magnetic levitation or "maglev". While no true antigravity has ever been discovered, scientists are searching for ways to simulate antigravity or levitation. Meanwhile, the only means people have of "countering gravity" is by going into space (weightlessness), use a plane called the Vomit Comet for momentary weightlessness, or employ maglev or other forms of levitation.

Electrogravity

Various thinkers among them Tesla tried to conceive a unified theory of electricity and gravity. They thought that gravity may somehow influence gravity or vice versa. Perhaps it was possible to make devices that could change from one energy to another. For the time being, electrogravity has gone largely nowhere and is now considered an issue of pseudoscience unless a breakthrough happens.

Forces Like Gravity

Gravity is a force known for pulling down. There are in physics discussion on "forces" that act like gravity in that they are able to cause a "pull". Many inventions have been made on these forces and more is being learned about them. Issues of such forces to explore are centrifugal force, Coriolis force, electric force, magnetic force, and so on.

Theories of Gravity (History of Gravity Theory)

Over the ages, many thinkers have sought to invent a theory of

gravity. The most famous ones known however are Aristotle's, Newton's, and Einstein's general relativity. Currently, Einstein's theory is the main theory of gravity in physics, but it is thought it will be overthrown or replaced some day. Various attempts to conceive a theory of gravity known to history are the following:
- Watt/Misner –Brans/Dicke/Jordan –LeSage –Fatio –Einstein/Cartan –Whitehead –Nordstrom –TeVeS –Einstein/Fokker –Ni theories –Yilmaz –Scalar/Tensor –Vector/Tensor –Scalar Field –Bimetric (Rosen) –Deser/Laurent –quantum gravity –loop quantum gravity –supergravity –Non Symmetric Gravity Theory –induced gravity –Horava/Lifschitz gravity

Gravity Probe A & B

These are the names of famous satellites sent into space. Their mission was to test for effects of gravity and predictions of general relativity.

"Hypotheses non fingo" (I frame no hypotheses!) -quote from Isaac Newton
- Regarded as a kind of confession that he did not understand what caused the force of gravity, hence he refrained from speculating on the nature of gravity other than to say it was a force.

In this section, we will examine a discussion on the nature of a force that holds the atomic nucleus together. It operates within the atomic nucleus and is considered the most powerful force known in Nature. For this it is given the name of the 'strong nuclear force'.

Strong Nuclear Force (Strong Force)

The Strong Nuclear Force

In the 20th century, physicists investigating the atom discovered the nucleus or "center of the atom". They discovered two unknown forces operating within the nucleus and they were called the strong and weak nuclear forces. The strong nuclear force is a force known to hold the atomic nucleus together. It is known protons have a positive charge and their "like charges" would cause protons to repel away from each other. This is thought to make a nucleus unstable and hence a nucleus would self destruct. The strong nuclear force has since then been a major topic of research. It is thought to be a very short range force operating only within the atomic nucleus, has immense energy, acts on particles called quarks, mesons, gluons, protons, and neutrons, and has other qualities. Its main theory is called by the long winded name of "quantum chromodynamics" or the quantum theory of color force changes. It is the force responsible for the atomic bomb and atomic reactors. Today, it hides many secrets still and efforts to unify it with other forces are underway.

Color Force

Inside the nucleus of the atom exist particles called quarks ("three quarks for muster mark"-Gell-Mann). Quarks are known to have a "force" called the color force mediated by "gluons". The word "color" does refer to the spectrum, but is "misused" to name a kind of version of the strong nuclear force. Its discussion tends to be technical and can be explored in books on QCD.

Nucleons

A nucleon is another name for a particle found in the atomic nucleus. Kinds of nucleon are the proton, neutron, meson, quark, and gluon.

Nuclear Energy

When an atomic nucleus is split, the strong nuclear force is "tapped" to release energy. This release of energy has been used in atomic bombs and nuclear power.

In this section, we will explore a discussion on a force that causes radioactivity. It too operates within the nucleus of the atom. Unlike the strong force which holds the nucleus together, this force is weaker compared to the strong force and causes some types of nuclei to self-destruct. For this it is given the name of the 'weak nuclear force'.

Weak Nuclear Force (Weak Force)

The Weak Nuclear Force

Another kind of force was also discovered in the nucleus. It is responsible for causing the nucleus to self destruct producing energy and particles and is called the weak nuclear force. The release of energy from the nucleus by the weak nuclear force is called radioactivity. Radioactivity is known to come in three types called alpha, beta, and gamma as famously discovered by the scientists, Marie and Pierre Curie and Henri Becquerel. The weak nuclear force is known to affect particles called W, Z, neutrinos, and so on. Today, the weak nuclear force has been "unified" with electromagnetism to produce a new fundamental force, the electroweak force. Research into this force continues with more surprises being found.

Radioactivity

About the year 1900, physicists named Marie and Pierre Curie, and Henri Becquerel investigated materials that glowed by a mysterious energy. This drama was later called radioactivity (or the active release of radiation). They went on to investigate new elements unknown to science later to be named polonium, radium, thorium, and so on. Eventually, it was discovered radioactivity came in three varieties called alpha, beta, and gamma. It was found ominously that exposure to radioactive elements caused cancer and other diseases leading to death. Since then radioactivity has been a major issue in physics leading to inventions like radium glow devices, atomic clocks, and so on. It was a fascinating issue and the Curies and Becquerel were given a Nobel Prize over this. Later on, Marie Curie was honored in the naming of the element curium and the Curie Institute.

FIELD AND FORCE

Please read about the issue of radioactivity and the careers of Becquerel and the Curies.

Marie Curie

Marie Curie is considered the foremost personality in the issue of radioactivity. She originates from Poland and moved to France. She married Pierre Curie and worked with Henri Becquerel. The Curies took to studying pitchblende (a substance) and discovered new elements called radium, actinium, and polonium. In time inventions were made that used radium's glow for use in watches and other tools. In time, Curie would be awarded two Nobel prizes and an institute would be founded dedicated to radioactivity (the Curie Institute). Curie went on to become a giant of physics history along with Einstein, Planck, and the quantum mechanics pioneers of the early 20th century. Today, Curie is honored in the element curium, statues, and history.

Ernest Rutherford

Rutherford was an English scientist from New Zealand. He lived about Einstein's time and performed the famous gold foil experiment. This experiment lead to the discovery of the atomic nucleus, neutron, and proton. Later on, Rutherford was awarded the Nobel prize, a laboratory was named for him, and an element called "rutherfordium" was named for him.

More Nuclear Forces

Physicists have learned about the strong and weak nuclear forces. However, there are controversies abounding that perhaps the nucleus hides other forces. For example, its thought there are particles called quarks, smaller particles than protons and neutrons. Quarks are themselves made of even smaller particles called preons, smallerons (smaller than quark), or smallestons (the smallest particle of all) as it has been speculated. Its thought these particles if they exist are held by unknown nuclear forces which can be called the preon force or smalleston force. Today, there is no evidence for either preon or preon force and hence this is a curious speculation only.

In the next section, we will explore a discussion on the nature of a force that is the union of electricity with magnetism. Exploration and innovation into this force has revolutionized civilization with inventions like radio, electronics, electrical power grids, and so on. It is today one of physics' classic issues and deeply studied topics. It is known by the name of 'electromagnetism'.

Electromagnetism

"Electromagnetism is the culmination of over 2000 years of Man's fascination with electricity and magnetism, two forces that he never understood till the time of Maxwell."

-class discussion

Electromagnetism (EM or electricity and magnetism as one)

In the 19th centuries, various thinkers like Michael Faraday, Oersted, and James Clerk Maxwell realized that two energies, electricity and magnetism were really aspects of a force called electromagnetism. For ages, it was thought electricity (the flow of charge) and magnetism (magnets) were separate and distinct forces. Maxwell is famous for proposing a theory stating that they were one force and he gave the world various physics laws called Maxwell's Laws. It was thought that such energies as light, heat, radio, X rays, and gamma rays were all products of an entity called the electromagnetic wave. This theory of Maxwell's later shook physics leading to many inventions. Today, electromagnetism or EM for short is a vital branch of physics filled with lots of research and offshoots. Inventions like electromagnets, X rays, and so on derive from this. Electromagnetism has been unified with the weak nuclear force as the electroweak force.

Maxwell's Laws

In the 19th century, James Clerk Maxwell would write down equations uniting the forces of electricity and magnetism. These equations would go on to become some of physics' most legendary statements describing the new force of electromagnetism or EM for short. They are complicated statements using calculus and higher math and have troubled physicists with their seeming easy and complexity ever since. Today, they are taught in virtually all electricity courses and the reader is encouraged to explore more definitive texts on these grand statements of science.

Faraday and Oersted

These two thinkers of the early 19th century would make discoveries that would lead to union of electricity and magnetism as EM. They contributed findings that would lead to modern appliances.

Particles of EM

Virtually all subatomic particles "feel" or "sense" for EM radiations in some way. Particles called the photon, electron, proton, neutron, quark, and so on all convey or "sense" for EM radiations in some way.

Electromagnetic Radiation (called EMR)

Electromagnetism was such a profound idea in physics that it lead to the discovery of an entity called the "electromagnetic wave", a result of the electromagnetic field. A particle of the EM wave is called the "photon" or particle of light. It was thought by adding or subtracting energy to the EM wave, various energies can be produced. A collected list of all known EM radiations is called the "electromagnetic spectrum". Physics knows of several kinds of EM energy like:

- Light. This is the most familiar energy and is the energy by which people see. It is known to be made of "colors" that go by the names of red, orange, yellow, green, indigo, blue, and violet (called ROY G BIV).
- Radio. This is an energy associated with radio stations. Famous names in it are Marconi, Hertz, Reber, Jansky, and so on. Its study is intense and more thorough books on It can be explored. Kinds of "radio wave" are TV waves, UHF, VHF, AM, FM, shortwave, and so on.

- Television waves. This is a class of radio waves that is used for television or TV broadcasts.
- Infrared rays. These are rays like heat used in astronomy and other areas.
- X rays. Discovered by Wilhelm Roentgen, this energy is used in medicine. X rays are known to come in "hard" and "soft" varieties and is known to penetrate substances resulting in X-ray photographs.
- High energy X rays. This is a kind of X ray known for very powerful bursts.
- Gamma rays. This is an energy known for being dangerous for causing cancer. It is also known for being a powerful kind of EM radiation emitted by radioactive elements and should be handled cautiously.
- Ultraviolet rays (UV rays). This is an energy more "powerful" than light known for causing sunburns and tanning. It factors in discussions on the "ozone hole".
- Cosmic rays. This is an energy coming from powerful stars in the universe and is a hot topic of exploration in physics research. They are intensely studied at the world famous Pierre Auger Observatory in South America.
- Ultra-high energy cosmic rays. This is class of cosmic rays known for exceptionally high energy that makes them seem distinct from ordinary cosmic rays. They are whimsically called "yoohekker rays" derived from the statement UHECR which stands for ultra high energy cosmic ray.
- Microwaves. This energy factors in microwave ovens and a type of laser called the "maser".
- Terahertz rays. This energy is a fairly unknown type of EM radiation.
- Long waves. This energy can be thought of as the weakest kind of EM wave or a "weak radio".
- Bremstrahlung. This refers to EM radiation brought about by electron interactions.

FIELD AND FORCE

- Sub-millimeter waves. These are EM waves of a wavelength smaller than a millimeter.
- Big Bang rays. These are EM waves emitted from the Big Bang (cosmic microwave background radiation). They are microwave in nature.

Uses of EM Radiation

Various thinkers have taken to inventing with EM radiations. They have invented devices that today are "classic tools" of physics. Presented here is a discussion on some tools and uses for EM.

- Radio. This refers to using radio waves for broadcasting, radar, and so on.
- Astronomies. This refers to devising telescopes to observe in a type of EMR, hence opening up a sub-branch of astronomy.
- Optics. This refers to the science of light and its uses.
- Lasers. This refers to beams of light and their uses. Other EM radiations can be used to make a "laser" of some kind. Lasers of other EM radiations are so named by replacing the "l" for light with the letter or representation of another kind of EM radiation. A microwave laser is hence named maser and an X-ray laser is called the Xaser or Xraser. Please now try to invent your own kind of "laser name".
- Lamps are devices that glow in a kind of EM radiation.
- Medicine uses EM for tools like laser surgery, MRI, PET scan, and CAT scan.
- X rays are used to see into living organisms in medicine.
- Gamma rays are used to sterilize against bacteria.
- Nightvision is a device that senses in heat when its dark out.
- Toasters are appliances that "toast" bread for food.
- Microscopes. This refers to "probes" to see small things in other kinds of EM radiation.
- Detectors. This refers to devices that sense for a kind of EM radiation.

- Ovens. These are devices that "cook" using a kind of EM radiation.
- Tanning. This refers to people getting a "sun tan" by exposure to ultraviolet radiation.
- Electronics. This is a branch of physics discussing devices like VCRs, DVDs, video games, TV sets, and so on. These machines use devices that harness EM for use in circuitry.
- Weapons like particle beams, ray guns, EMP, and so on all use EM for warfare.
- Strobe is a kind of lamp used in discos.

There are in reality many uses for EM radiations. The reader is encouraged to explore the various uses and try to innovate yet still original applications for EM radiations.

The Dangers of EM Radiations

Its known that exposure to various kinds of EM radiation can have terrible effects on living things. Its known that UV rays cause skin to tan, but it can also cause sunburns. X-rays are useful in medicine, but over exposure to them can cause cancer and other ailments. Gamma rays are known to cause cancer and direct exposure to them should be avoided. Cosmic rays are just too dangerous to be exposed to and can cause death without protection.

Thinkers of Electromagnetism

EM has a long, storied history with people making epic contributions to its understanding. Today, physics books celebrate these people as its pioneers. It is recommended to the reader to explore their lives closer. Examples of such names are:
- Maxwell (Maxwell's Laws) –Oersted –Ampere (Ampere's Law) –Romagnosi –Faraday (Faraday's law) –Heaviside –Coulomb (Coulomb's law) –Ohm (Ohm's law) –Volta (voltage) –Kirchoff (KVL, KCL rules) –Benjamin Franklin (kite experiment) –Poincare –Lorentz (Lorentz Transformation) – Millikan (oil drop experiment) –Planck (blackbody) –Einstein

FIELD AND FORCE

(photoelectric effect) –Edison (direct current) –Tesla (alternating current) –Weber –Henry –Feynman, Schwinger, and Tomonaga (QED) –Reber –Jansky (radio astronomy) – Weinberg, Salam, and Glashow (electroweak force) –Thevenin (circuits) –Gauss (Gauss' law) –Biot and Savart (Biot-Savart law) –Debye –Giauque –Joule –Hannes Alfven (ambiplasma) –Maricourt –Gilbert –Otto von Guericke –Galvani -CF duFay

Ideas of Electromagnetism

EM today is a complex subject taught in higher level physics. It takes years to learn EM and its knowledge is used to train electrical engineers, electricians, and other professionals. Provided here is a list of ideas to explore in learning EM:

- electric field –magnetic field –electromagnetic wave –photon –charge –magnetic pole –electron –proton –neutron –current –resistance –voltage –conductance –capacitance –inductance –elastance –power –work –electromotive force –resistivity – permittivity –susceptance –conductivity –superconductivity –semiconductors –insulators –magnetic flux –electric flux –magnetic moment –magnetostriction –magneton –coercivity –remanence –flux –battery –anode –cathode –ion

Uses of Electromagnetism

EM is today one of the most widely harnessed energies around. Inventors have made all kinds of inventions from EM. Examples are lasers (light rays), microwave ovens, telescopes, microscopes, radio devices, telephones, detection devices, counters, military weapons, TV sets, computers, video games, and so on. EM technology can be said to have revolutionized civilization with all new kinds of invention. Today, new inventors are hard at work trying to find new uses with EM. Major thinkers into this area are Thomas Edison and Nikola Tesla. These men are famous for beginning such things as the light bulb, electrical technology, and so on. Please read biographies on this pair of inventors.

Radio

Radio was discovered by men like Marconi, Hertz, and others. It is a kind of EM radiation of weaker energy than light. It has been found that it can be used effectively in communication and other offshoots. Discussions on its offshoots are:

- Radio stations. Inventors took to broadcasting messages with radio. This lead to the creation of installations that broadcast radio to the public. This resulted in such things as rock stations, talk radio, public radio, the president's radio address, music stations, and so much more. Today, it's a vibrant medium for people who like to listen to music or explore careers as disc jockeys, shock jocks, talk radio commentators, and so on.
- Radar. This is a word that means "radio detecting and ranging". It is a means to use radio to locate objects.
- Ground penetrating radar. This is an offshoot of radar to locate objects in the ground.
- Radio astronomy. Inventors named Reber and Jansky developed a "radio telescope" that opened a new branch of astronomy. Installations using them are called Arecibo, VLA, Jodrell Bank, Green Bank, and so on.
- SETI research. Here astronomers search for signs of alien intelligence by radio.
- Ham radio. This is a hobbyist activity of using radio to communicate globally.
- Walkie-talkies. These are communication devices using radio.

There are in fact many kinds of radio technology. Please study the many kinds.

X Rays

Wilhelm Roentgen discovered X rays in the 19th century and was awarded a Nobel prize over this. X rays were found to pass through living things and could be photographed showing the interior of bodies. Since then, X rays have changed medicine and are now used routinely in medical practices worldwide. X rays have been used to make lasers, give "Superman" his "supervision", sterilize against bacteria, and do so much more. It was a great discovery in physics history.

Microwaves

Microwaves are a kind of EM radiation weaker in energy than light. Inventors have investigated it and have devised inventions like the "maser" (microwave laser) and microwave oven. Today, it's a heavily investigated energy used in astronomy and technology.

Charge

Charge is the notion of an energy presence in space. It is "emitted" from point charges, electrons, protons, and quarks in an energy presence called an "electric field". Charge comes in two types of "positive" and "negative". When two particles of "like charge" or the same charge meet each other, a force is caused to "push" the particles away from each other or repulsion. When "unlike charged particles" meet each other, a force is caused to "attract" each other. Charge is a fundamental issue in physics and is basically of unknown origin. It is a quality of particles that allows for electricity, lightning, static electricity, sparks, and other electrical phenomena. A subatomic particle called the "quark" is known to have fractional charge or charge of smaller units than protons or electrons.

Electromagnetic Induction

In 1831, both Michael Faraday and Joseph Henry carried out experiments in electricity. They found that when they were conducting changing magnetic fields that an electric current would be aroused. From this, various kinds of invention were derived that would lead to today's modern electrical equipment like transformers, power supplies, electric wires, dynamos, batteries, and other machinery.

Electromagnetism is known as the united force of electricity with magnetism. Both these forces were historically thought fundamental in nature until they were found united as electromagnetism. In the following sections, we briefly review these 'sub-forces' in electricity and magnetism.

THE GRAND SCIENCE OF PHYSICS

Electricity

"Electricity was one of the greatest discoveries of history like fire, the wheel, language, and science. It revolutionized human existence and gave the world modern technology."
-class discussion

Electricity (Elektros)

For centuries, electricity was thought to be its own fundamental force. It factors in such things as lightning, ball lightning, static electricity, the electric field, current, and so on. It comes in two varieties called positive and negative charge. Its known that particles can carry either type of charge. Particles that carry unlike charges attract each other and particles that carry like charges repel each other. Particles known to carry charge are called the proton, electron, antiproton, and positron. Quarks are particles thought to carry "fractional charge" (or charge of smaller amounts). When a particle with a charge "flows" in a wire, it is said to be conducting (current) producing electricity. Electricity has been investigated for centuries and has lead to modern technology like powerlines and appliances. A primitive invention called the Leyden Jar is famously known to be one of the first kinds of electric invention. There is a famous controversy in physics over the nature of current. Current is the flow of electrons in a wire. There are two kinds of current in direct and alternating varieties. A famous controversy has the two inventors Tesla and Edison championing one of the currents for modern technology. Tesla won this battle and it is known historically as the "war of the currents". Research into it continues with more secrets being found. Please read textbooks on this interesting kind of energy.

Thinkers of Electricity

Electricity was for many ages a baffling mystery. So many thinkers sought to unravel its secrets and today's understanding of electricity is indebted to the people who helped solve its mysteries. Presented here

is a list of people who all contributed in some way to unraveling the secrets of electricity:
- Ohm (Ohm's law) –Franklin (kite experiment) –Volta (voltage) –Galvani (galvanometer) –Kirchoff (Kirchoff's laws) –Weber –Oersted –Faraday (farad) –Edison –Tesla –Ampere –Gauss –Maxwell –Thomson –Alfven –Thevenin –Henry –Heaviside –ancient thinkers

Amber, Lightning

The ancients did not understand what electricity was. They realized that by rubbing a piece of amber (stone) with wool then sparks would be produced. They also knew that leaving cloth sitting around would produce something later to be called "static electricity". Also, it was believed that perhaps sparks and lightning were related, but this suspicion went unrealized until Benjamin Franklin performed his famed kite experiment.

Forms of Electricity

Electricity has been investigated for centuries. It was in time realized that lightning, sparks, static electricity, aurora borealis(Northern Lights), aurora australis (Southern Lights), ball lightning, van der Waals forces, plasma displays, solar wind, and so on were all aspects of electricity.

Lightning

Its known in thunderstorms that there will be epic displays called lightning. Lightning is an electrical disturbance caused by irregular amounts of charge accumulating in the ground and in the clouds. Lightning is infamous for killing people by electrocution, causing fires, and being an uncontrollable force of Nature.

Benjamin Franklin

Franklin was a leading statesman and thinker during the American Revolution. He took interest in electricity and decided to perform an experiment. He made a kite with a long string with a key attached to it. He went flying the kite during a thunderstorm. He believed that that lightning was electricity and would somehow cause an electric flow from the kite to the key. This he did to prove lightning was electricity. He is also famous for naming the types of electric charge positive and negative as well.

Electric Field

It was realized by thinkers that various objects radiate an energy presence that allows electricity to occur. Eventually, this lead to the idea of the electric field or "energy presence of electricity in space". A charged particle that moves in an electric field is said to "feel" an electric force (a push or pull caused by an electric field).

Coulomb's Law

Charles Agustin de Coulomb ("koo lomb") was a 19th century thinker into electricity. He discovered an equation relating "electrical charges" to electric force. Today, this law and its equation are studied throughout physics and is a central teaching in electricity. It is here expressed by the equation: $F=kQq/r^2$.

Concepts of Electricity

Electricity has gone on to become one of physics' most heavily studied energies. Various thinkers have created ideas of how to understand electricity. Presented here is a list of the various ideas of electricity.
- electric flux –conservation of charge –positive charge –negative charge –flow –electric potential –Gauss' law –solenoid –resistance –conductivity –conductance –resistivity –current –Ohm's law –voltage –electric field –electron –proton –neutron –capacitance –inductance –remanence –susceptance –field lines –point charge –dipole –quadrupole –superposition –Coulomb's law –electric force –permittivity –resistor –semiconductor –conductor –center of symmetry –Gaussian surface –electrostatics –electrodynamics

FIELD AND FORCE

Please study the myriad concepts of electricity.

Cathode Rays

In the 19th century, various thinkers like Kirchoff, Crookes, Thomson, and others studied a mysterious radiation called "cathode rays" in a device called the Crookes Tube. In time they would go on to discover the electron and realize that cathode rays were a kind of electricity. They would in time devise a machine called the "cathode ray tube" or CRT. This device would later be used to invent the television set, the oscilloscope, the computer monitor, and the vacuum tube.

Philo T. Farnsworth

Farnsworth is a thinker credited with inventing television. Television (TV) has since gone on to be a key invention of history. It lead to video games, game shows, TV news, cable TV, computer monitors, Hollywood, Bollywood, and many offshoots.

Electron ("Electron flow is electricity")

Various thinkers in the 19th century (JJ Thomson, Kirchoff) were exploring electricity's nature. Some went on to report the discovery of a particle that seemed to convey electricity, hence it was named the electron. They found that when the electron "moved", it created an energy disturbance that caused light, heat, and so on and was named the "electric current". Today the electric current is harnessed in inventions from light bulbs to toasters making electronic devices possible. It was found the electron "orbited" a mass object called the "nucleus" and hence this formed a model of the atom. Today, electrons are one of physics' most heavily studied particles used in particle accelerators and other devices.

Electrostatics and Electrodynamics

The study of electricity has resulted in the creation of two of physics' sub-sciences. Electrostatics is the study of charges that do not move. Electrodynamics is the study of charges that move or are in motion. They were invented to study how electricity behaves and are continuing evolving even now. Later thinkers would apply quantum mechanics to the study of electrodynamics. They would derive a new branch of physics called by the long-winded name of "quantum electrodynamics" or QED. Its main thinkers are people named Feynman, Schwinger, Tomonaga, and Dyson. The first three were jointly awarded a Nobel Prize over its creation.

Superconductivity

Its known when masses are cooled to temperatures near absolute zero, they change behavior as to how they allow electricity to flow. In ordinary "conductors" (substances that allow electricity to flow), electrons will collide with matter producing heat and light in what is called "resistance". Near absolute zero, matter will allow electricity to flow where there is no resistance and this is called "superconductivity". This effect of matter was originally discovered by Heike Kamerlingh Onnes and he won a Nobel Prize over it. Since then, superconductivity has fascinated physicists as they wanted to understand and control it. Three thinkers named Bardeen, Cooper, and Schrieffer composed a theory on it called the BCS theory "explaining" superconductors. Eversince, physicists have wanted to make superconductors that could "superconduct" at temperatures like "room temperature". Its thought if they could make such a substance, then this could start a revolution in "engineering" and would require a new theory of this issue. For the time being, this issue has won its share of Nobel Prizes and physicists are still continuing to dream of making a "room temperature superconductor".

Uses of Electricity

Electricity has today become modern civilization's most used energy. It has found use in such inventions as the following:
- Light bulb. Edison realized that if an electric current were passed through a wire in a gas container, a glow would result. This lead him to invent the first light bulb which changed society immensely.

FIELD AND FORCE

- Stoves and ovens. Inventors realized that by passing an electric current through wiring, heat was able to be given off. This lead to inventions like stoves, ovens, warmers, toasters, and indoor heating appliances.
- Telephone. Alexander Graham Bell realized an electric signal can convey messages. This was used to invent the telephone and earlier thinkers invented the telegraph over this as well.
- Video game. This is a "contraption" to play games on television sets.
- Robots. These are "contraptions" to work, mimic Man, and perform for use.
- Internet. This is a collected network of computers useful in society.
- Computers. Various thinkers realized electric signals can be used to calculate and convey messages. Later this found use in the invention of calculators and computers.
- Circuits. Various thinkers realized that batteries could produce electric currents. This lead inventors to devise constructions called "circuits". Circuits have qualities like power supplies, switches, resistors, diodes, capacitors, and other device that exploit the nature of electricity for machine use.
- Lightning rod. This is a stick placed on buildings made of metal. Lightning will strike it and conduct electricity to the ground. This was invented to prevent lightning from causing fires on wooden buildings.
- Television. Also called the cathode ray tube, this device allows for moving images. It went on to revolutionize much of the world in communication and entertainment.
- Electric chair. Various thinkers were searching for new and human ways to execute criminals. They built a chair and condemned prisoners were made to sit in "the chair". An electric shock would be passed through the criminal killing him "instantly".

- Particle accelerator. Here electricity is used to move particles to collide with a target hopefully to produce other kinds of subatomic particle.
- Electroscope. This is a device that detects the presence of an electric charge.
- Appliances. Electricity has been used to make video games, TVs, radio, computers, DVD, lights, telephones, compact discs, toasters, transistors, and other devices.
- Electromagnet. This is a device where electricity is passed through an iron core to cause a magnetic effect. It is used heavily in metal recycling plants.
- Dielectrics. This refers to materials that have induced dipole moments.
- Meter devices. This refers to various tools to measure electricity called by such names as ammeter, voltmeter, galvanometer, and so on.
- Electrified fence. Various inventors have built fences about buildings. Electricity is then passed through these fences and they act as security barriers against intruders.
- Electrolysis. Electricity is passed through substances like water. Water is broken up (or dissociates) into the gases oxygen and hydrogen.
- Electroplating. This is the use of electricity to "coat" an object in gold, silver, or other metal or substance.
- Electrification. This is the activity of making a place having electrical technology.
- Motor. These are devices that use electricity to cause motion. They are used to power cars, motorcycles, ships, and other devices.
- Generators. These are devices that use some form of power to generate electricity. Devices like this are found in hydroelectric dams, geothermal plants, nuclear power plants, and so on.

Please read about the many inventions that discuss the use of electricity.

FIELD AND FORCE

Controlling Electron Flow (Current)

Its been found by many inventors that electron flow (or electricity) can be controlled for use in many inventions. Discussion on this goes as follows:

- Electron flow must be caused by some power supply (battery or source).
- It flows through wires extending from the power source.
- It flows whereby it seeks a state of uselessness (called the ground).
- It flows through devices where it can be used in some way.
- It flows in devices where it is resisted of flow (resistor). Resistance can be used to cause heat (toaster), light (light bulb), or other effect.
- Switches can block or allow electron flow.
- Capacitors build up electron flow where a spark is emitted later on as the electron flow crosses the gap between plates.
- Inductors use electron flow to generate magnetic fields.
- Lamps use electron flow to generate light.
- Various electronic devices like LEDs, thyristors, memristors, diodes, logic gates, telephones, mathematical circuits, and so on use electron flow for their purposes.
- Insulators attempt to block electron flow.
- Circuit breakers are devices that shut down circuits.

Kirchoff's Laws

Gustav Kirchoff was a 19th century thinker in electricity. He devised laws by which electric circuits can be so controlled. His two most famous laws are called Kirchoff's voltage law (KVL) and Kirchoff's current law (KCL). They are technical discussions explained in electricity courses.

Generating Electricity

Inventors for ages have tried to generate electricity by many methods. Presented here is a discussion on some methods known.

- Piezoelectricity. This is the generation of electricity by "squeezing quartz".
- Thermoelectricity. This is electricity by heat.
- Photoelectricity. This is electricity by light and was investigated by Einstein.
- Hydroelectricity. This is electricity by water flow by use of dams.
- Pyroelectricity. This is electricity produced by fire or similar effects.

Electricity can be generated by most any kind of power in Nature (solar, wind, current, geothermal, battery, fuel cell, etc.). It requires that a powersource can generate energy by which it can be changed into electricity. Inventors are always searching for interesting and novel ways to generate electricity.

Electronics

Electronics refers to the technology of electricity. It discusses such devices as TVs, radio, computers, video games, DVD, VCRs, telephones, music players, and so much more. It is a complicated subject taught in connection with electricity. People skilled in electronics can go on to careers as repairmen, inventors, installers, etc.

Electrical Engineering

Electricity has come along way since the days of Franklin's kite experiment. It has now grown so bewildering and complicated that it has lead to the creation of a branch of engineering dedicated to electricity. People today are trained in college to handle all manner of electrical technology like transformers, wires, circuit breakers, ammeters, voltmeters, and so on. Many electrical engineers are trained to be repairman, installers, electricians, and so on. It is today a lively branch and offshoot of electrical technology and education.

Ions

Ions are atoms that have gained or lost electrons and thus have become "charged". Ions are studied in particle physics and are used in chemistry. Research into them is lively with new discoveries being found.

Electricity has gone on to be modern society's most used energy. It is vital from everything from lighting to powering appliances. It seems almost inconceivable that people once lived without electricity, but today it is everywhere used for almost everything.

> "Lightning was thought to be the power of God and thunder was the loudness of His voice in moving the wind. It is an incredible display of might and fury that awed the ancients and impresses scientists."
>
> -class discussion comment

In the next section, we will explore a discussion on a counterpart to electricity, electricity's "twin" as an energy of electromagnetism, magnetism.

Magnetism

Magnetism (B force)

For centuries, thinkers have thought magnetism was a fundamental force and is produced by a mineral called lodestone. It is known to come in its own type of charge called the "magnetic pole". It is said there are North and South magnetic poles and the planet Earth is known to have a pair. It is known a North Pole will attract a South Pole, but two North and two South Poles will repel each other. Magnetism is basically found in metals like iron, nickel, and steel. Objects that possess magnetism are said to be magnets and there are many kinds like electromagnets, horseshoe magnets, and bar magnets. Its widely thought a magnet of only one pole is possible and it is called the magnetic monopole (one pole). However, searches to find it have found nothing. Magnetism factors in many kinds of technology like computer disks, toys, compasses, and so on. Research into it is lively with many more secrets being discovered. Please read textbooks on this interesting kind of energy.

Concepts of Magnetism

Magnetism like its "twin energy" of electricity has been intensely studied. It has lead many thinkers to derive ideas on how to think about it. A compiled list of its many basic ideas are the following:

- magnetic field –magnetic force –magnetic flux –North magnetic pole –South magnetic pole –electromagnet –lines of force –magnetic axis –Van Allen belt –magnetosphere –magnetic levitation –fusion –Hall effect –magnetic domain –hysteresis –magnetic dipole –paramagnetism –diamagnetism -ferromagnetism

Thinkers of Magnetism

Magnetism has been mysterious for ages driving many thinkers to learn its secrets. Presented here is a list of people who in some way contributed to unraveling the secrets of magnetism:

- Hall –Blas Cabrera –Kirchoff –Oersted –Volta –Ampere –Faraday –Galvani –Maxwell –Gauss –Alfven –Biot and Savart –Coulomb –Henry –Lenz –Gilbert –many thinkers

Lodestone (Magnetite)

This is a mineral known to ancient thinkers. It is essentially iron ore and has magnetic effects. The ancients experimented with lodestone noticing its curious effects, but were unable in their time of superstition to explore its nature further. It was left for thinkers of later ages to discover magnetism which originated with lodestone.

Magnetic Field

Various thinkers into magnetism realized that magnets have two poles (or kinds of magnetic charge) dubbed North Pole and South Pole. Magnetic poles seem to radiate "lines of force" away from each other and leading to another pole. Magnetic lines of force illustrate the existence of the magnetic energy presence called the magnetic field.

Kinds of Magnetism

Magnetism is known to occur with metals like iron, nickel, steel, and so on. Its been discovered magnetism comes in types called ferrimagnetism (of iron), antiferrimagnetism, paramagnetism, and so on. Discussion on them tends to be technical and can be explored in more definitive books on magnetism.

Geodynamo

Its known that the Earth has a molten core of iron. This core of iron creates the Earth's magnetic field. This magnetic field is known to trap particles from the Sun and hence creates a region of space called the "magnetosphere". Its known that the Earth's magnetic poles sometimes change orientation where North Pole becomes South Pole and South Pole becomes North Pole. Its thought such "magnetic reversals" cause extinction events and other calamities. There is an effort to try to harness the "geodynamo" for a source of energy (sometimes called Earth energy).

Compass

In ancient China, various thinkers took to realizing that lodestone could be used to find direction, this lead to the development of the compass. Since then the compass has become an indispensable tool used in navigation, camping, and other uses.

William Gilbert

This man is an ancient author who wrote the classic text, De Magnete discussing magnetism from ancient times. Please take time to read about the many classic texts on magnetism to fully explore magnetism's nature.

Electromagnet

It has been found by inventors that if you pass an electric current through iron materials, a magnet can be created. This magnet is also called the electromagnet. Electromagnets have been used from everything from particle accelerators to car crushing operations.

Magnetronics

Magnetism can be harnessed to form electronic devices. A branch of physics that studies them is called magnetronics.

Uses of Magnetism

Magnetism has been regarded as one of physics' greatest discoveries. Inventors have used magnetism in the following inventions:

- wall magnet –compass –electromagnet –toys –cars –car crushing –metal recycling –fusion research –particle accelerator –curios –appliances –chemistry equipment –bar magnet –horseshoe magnet –magnetic levitation –magnetic seal –magnetic circuits –magnetohydrodynamics (Hannes Alfven) –magnetic loops – magnetooptics -solenoid

Please read about the many uses of magnetism.

Magnetic Monopole

There is a speculation in physics that perhaps a lone magnetic pole particle called "magnetic monopole" exists. Various people have sought after it including a scientist called Blas Cabrera (famously shown not to have found it), but so far nothing has ever been found. Its thought such particles may exist, may not exist, or that one lone one exists in the universe. For now, no evidence of its existence has ever appeared, but its issue is tantalizing in physics today.

Fusion

There is a worldwide quest in physics to harness the power of the Sun or fusion. They want to make a reactor that can fuse hydrogen to make helium and energy. So far the technical challenge to do so is very tough. Various thinkers have devised "fusion reactors" to use magnetic fields to contain "plasmas" to make fusion happen. So far research into this has not worked, but efforts continue. In Russia, an experimental fusion reactor called the tokamak has been made trying to exploit magnetism for fusion.

Magnetism is today a well understood energy. It has been exploited for countless inventions and even now more kinds of invention are being found. It started off as an anomaly in lodestone to the ancients and has evolved since to become one of the modern worlds most recognized energies.

"Lightning for centuries was thought to be a demonstration by God of His immense power over Man."

-Voigt

In the next sub-section on energy, we will explore the discussion of physics' newest fundamental force. It involves the unification (or making one) of the weak nuclear force and electromagnetism. Physics has known for years that there were four fundamental forces. Because of the discovery of the electroweak force, it can be said that there are just now three fundamental forces.

Electroweak Force

The Electroweak Force
(or Electromagnetweak Force, dubbed EW or EBW)

In the 1960s, three physicists called Sheldon Glashow, Steven Weinberg, and Abdus Salam proposed a theory that EM and the weak nuclear force were the result of one force dubbed the electroweak force or Emweak force. This was a major achievement in physics that lead to this trio receiving the Nobel prize in 1979. It is conveyed by particles called the W and Z bosons discovered by the Nobel prize winning scientists Simon van der Meer and Carlo Rubbia. It represents the first discovery of a genuine new force of Nature since the nuclear forces in the early 20th century. Today, the electroweak force is a major topic of research. As of now, it can be argued there are in fact just three fundamental forces (gravity, electroweak, and strong nuclear force). There are attempts to unify all three of these forces, but today success has not been achieved.

EW's Nature

The electroweak force is known to have three qualities to its nature. It can change flavor (a subatomic quality), conserve parity, and is "mediated" by massive gauge bosons (particles).

EW's Discovery

EW was found to exist by a 1973 Gargamelle (particle accelerator) discovery of neutral currents and the 1983 discovery of the W and Z bosons.

Higgs Mechanism

The Higgs Mechanism (Peter Higgs) is a discussion on how particles acquire mass. This has lead to the speculation that a particle called the Higgs boson exists, but as of now it has not been discovered.

EW Thinkers

The discovery of EW was a major event in physics. It was a major step in the drive to unify all forces as an aspect of one force. In some sense, it moved physicists closer to finding the grand unified theory. Its major thinkers go by the names of Salam, Weinberg, and Glashow and were all awarded the Nobel prize over this grand achievement. Salam would found a physics institute in Italy. Weinberg would author the classic book, The First Three Minutes and become a modern physics guru. Glashow would go on to write books and become a titan of late 20th century physics. Their work lead to the discovery of the W and Z bosons thought to convey the EW force. The discovery of these particles at the physics institute called CERN was also a major event

FIELD AND FORCE

in physics. This drama won its thinkers Van der Meer and Rubbia the Nobel Prize.

Please explore the lives and careers of the men who would pioneer the electroweak force and its discussion.

Strongelectroweak (Grand Unification Theory)

This word refers to attempts to unify the electroweak force with the strong nuclear force. So far research here has gone largely nowhere.

Gravitoelectroweak

This word refers to attempts to unify the electroweak force with gravity. Like the above, it too has gone largely nowhere in research.

Other Forces

There are controversies in physics that perhaps other forces exist in addition to the nuclear forces, gravity, EM, and electroweak force. As of now, these controversies are just problems and none have been confirmed to exist. However, whenever a new force is discovered, it is genuinely regarded as a major event of physics history. Examples of such controversies to explore are:

- hypercharge –preonic force –antigravity (repulsity) –Pioneer force (a controversy over what is disturbing the Pioneer 11 space probe) –extraweak force –tachyonic force (force of tachyons) –inflation force (inflaton field) –the grand unified force (or superforce) –electroweakstrong force –dark energy –phantom energy –chi force or energy –infinite energy force –fifth force –sixth force –seventh force –pyschic force –mind over matter force –cosmic acceleration force –orgone force –Mach force (Mach's Principle) –smalleston force –extra nuclear forces –quantum gravity force

The electroweak force was a fantastic drama in physics. It began with thinkers like Salam, Glashow, and Weinberg searching for ways to unify the forces of Nature. Today, it is a largely obscure subject found mostly in particle physics.

"In philosophy, there is talk of the One and the Many, the Whole and the Parts. Man studies the many and the parts in science, but he wishes to know the One and the Whole. He cannot know the One and the Whole and that has been the dream of thinkers for ages."

-Voigt

In the next section, we will examine discussions on the immensity of physics. We will review brief discussions on offshoots, curious issues, sub-disciplines, and related subjects of physics. Physics is an immense science filled with a lot of technicalities, findings, branches, paths of research, and curious sub-issues. To better understand the fantastic universe of physics, we will examine sub-disciplines of physics.

Chapter 5
Disciplines of Physics

"A tree springs from a seed. Its roots extend into the ground to be nurtured by the ground, soil, water, and nutrients. The tree will grow into something grand with many branches. No one knows all the branches that will grow from a tree."
-physics lecture

Branches, Sub-Disciplines, and Offshoots of Physics

Physics is an immense science, its knowledge and education is vital to so many other subjects. There exist a vast body of other sciences and subjects that share in physics. Presented here are basic descriptions of various kinds of subjects physics finds itself involved with. Many subjects arise out of other sciences, yet they involve physics study in some way. Physics is more than just a study in equations and principles, it is a body of knowledge that affects so many other subjects. To really understand the enormity of the subject, it should be known that physics affects the modern world in ways so profound that no one can understand its true effects. Its offshoots are a vast labyrinth and there is no one alive who is studied in it all. Here are some descriptions of some subjects that physics has found a place in or requires its education to understand.

Magnetohydrodynamics

Nobel prizewinner Hannes Alfven would study how electricity could be generated by magnetic fields.

THE GRAND SCIENCE OF PHYSICS

Richard Feynman

Feynman would be born in the early 20th century and major in physics in college. He would live in a time where he would work on the atom bomb, work under Einstein, and live a monumental life in physics. He would write widely and win the Nobel Prize in 1965. He would live a life shaping physics with his writings, books, explorations in physics, and his comedian-like personality as well. He would become a fad in physics fascinating legions with his views and works.

Lunar Theory

This is a subject that speculates about the movements, the origin, and end of the Moon, Earth's natural satellite.

Physics Television

Physics is an immense and complex science. Its teachings pervade so many other sciences and topics. For this people would make TV programs trying to teach physics for the enjoyment of the viewing public. Many TV shows were made to teach physics to people who would not learn physics. Many of these programs would become 'classics' of science teaching and resonate still for their hosts, teachings, special effects, and more. Famous shows that would instruct in physics to explore are the following:

Steven Hawking's Universe
The Astronomers
Carl Sagan's COSMOS
NOVA
Bill Nye, the Science Guy
The Universe

Inventing

This is the activity of conceiving new ideas for technology or other applications. New ideas are then made into products to sell, books to write, and so on. New ideas can be protected by law by a document called a "patent" or "copywrite". Many nations maintain a patent

BRANCHES, SUB-DISCIPLINES, AND OFFSHOOTS OF PHYSICS

office. Inventing can be explored by studying creativity methods, great inventors like Edison, Tesla, and others, and many books. Its activity is age old and worldwide and many people take up inventing as a hobby. Often many physicists find alternative careers as inventors. Famous inventions of history are fire, the wheel, the airplane, computer, and much more. Inventing profoundly affects the world in inventions, gizmos, fads, tools, appliances, and in many domains. It is a vast world all its own with sub-issues like:

Patent Examination
Creativity Research
Robotics
Popular Science, Popular Technology
Invention Companies

Great Inventors

Many people would define inventing with machines, devices, innovations, and in many ways. There work pervades the modern world in gizmos, developments, and many achievements. Some inventors would become so phenomenal as to become famous in history. Such inventors would be studied for their ideas, innovations, suggestions, comments, and in other domains. Famous names in this are the following:

- Thomas Edison -Nikola Tesla -Steve Jobs -Leonardo da Vinci -Heron of Alexandria -Archimedes -Aristotle -Alexander Graham Bell -Yoshiro Nakamatsu
- Einstein -JR Oppenheimer -Wright Brothers -Sikorsky -Jacques Cousteau -Benjamin Franklin -Alfred Nobel

Tesla Inventions

Nikola Tesla was an inventor in the late 19th/early 20th century. He was born in the European nation of Serbia and would work under Thomas Edison. He was famous for pioneering electrical engineering and other devices. He would become something of an anomaly of an inventor famous for work in electricity, radio, and other domains. He was often mentioned as a future Nobel prizewinner, but missed out on this award. He would become a titanic name in inventing known for the Tesla coil and many fantastic inventions. He would conceive so many ideas, that intrigues would begin to find out just what he knew. He was a curious personality known for a rare genius in inventing. Many books would be written on him and he still fascinates inventors to this day.

Nakamatsu Discussion

In Japan, there lives the inventor named Yoshiro Nakamatsu (or "Dr. Nakamats"). He is described as one of history's great inventors for his epic achievements. This subject is an analysis of his life, work, and legend.

Patent Law

This is a branch of the legal profession that studies and prosecutes for violations of patents. Lawyers versed in this area bring lawsuits against companies and "entities" accusing them of "stealing" ideas on a patent. Many physics-educated people find a new profession in this domain.

Patent Conflicts Discussion

This is a subject dealing with controversies about inventors having invented the same thing about the same time. It deals with issues of priority, claims of ownership of a patent, and so on.

Military Science

This is a vast domain of subjects into studying sciences concerned with the military. Physics would intertwine deeply with these subjects and define them as well. Overall military science is taught in war colleges and on military bases. Sub-domains of its study are:
- bomb physics -gun science -special forces -nature of war -war games -naval engineering -explosives -weapons -strategy and tactics

BRANCHES, SUB-DISCIPLINES, AND OFFSHOOTS OF PHYSICS

Nuclear Plant Safety

Its known that nuclear power plants are highly dangerous for their radioactivity. They are elaborately designed and maintained securely so that no radioactivity leaks into the environment. Chernobyl in Ukraine is considered the worst nuclear accident in history and is a lesson in the dangers of places like this. Topics here include:

- the China Syndrome -Three Mile Island -Chernobyl -nuclear accidents -meltdown -nuclear waste -accidental release -exposure

Environmentalism

This is a movement to arouse interest in pollution, whaling, overhunting, and other environmental problems. Environmentalists seek to solve these problems, protest, and get governments to be responsible with Nature. Famous groups here are Greenpeace, the EPA, Friends of Earth, and many other organizations.

Creation Mythology

Many ancient societies and primitive tribes would have their own stories of how creation came to be. Many stories would discuss a primordial reality, a creator god, the gods, and the use of magic to create the world. Many of these stories would prove fanciful and are now discussed in physics primarily as a recreation.

Toxicology

This is a science into the nature of poisons, toxins, and harmful substances. In physics, types of element and substance are known for poisonous or radioactive effects harmful to life. Topics of discussion are:

- radon -plutonium -nuclear waste -asbestos -lead -snake toxin -uranium -radium -radioactive elements

Oncology

This is the study of cancer and its nature, a branch of medicine.

THE GRAND SCIENCE OF PHYSICS

Medical researchers have taken to trying to use X-rays, chemicals, and physical effects to destroy cancer cells, hopefully finding a cure for cancer.

Boscovich, Dalton Study

Boscovich and Dalton are two epic thinkers into atoms from previous centuries. They are studied of their views and influence upon physics.

Alhacen Study

Alhacen is an Islamic thinker from the Middle Ages famous for his work on optics. His books still exists and he is studied in physics still. Other thinkers to explore are Avicenna and Averroes. In the Middle Ages, Arab scholars would found the epic 'Islamic Golden Age' where education would flourish in Arab nations during this time.

Physics Education

Physics is an immense and complex science. It has fascinated many to take up teaching it. For this many would explore ways to teach physics in ways like:

- physics camps -physics television -physics museums -displays -writing -demonstrations -distributing literature -outreach efforts

Cryptozoology

This subject literally meaning "the study of mysterious animals" has been practiced for ages. It concerns exploring such issues as Bigfoot, the Yeti, the Loch Ness Monster, the chupacabra, the thunderbird, Ogopogo, fairies, dragons, and so on. Various people have researched, lead expeditions, and delved into exploring whether these 'animals' were real or not. While some like the Vu Quang Ox and coelacanth (a prehistoric fish) have been discovered, many have persistently remained unknown. Bigfoot is the mystery of a "hairy man creature" wandering the woods of the United States and the yeti is a counterpart "creature" supposedly wandering the Himalaya Mountains. Many have sought for these creatures, but none have ever been proved to exist despite "evidence" like footprints, hairs, dung, and innumerable sightings. This subject is for those who like wondering whether or not these animals exist and is considered a subject of pseudoscience and zoology.

Bigfoot Research

In the Pacific Northwest of the USA there exists a legend of a hairy manbeast stalking the forests here. It is called by such names as Bigfoot, Sasquatch, Momo, the Skunk Ape, Omah, Mountain Devil, and so on. Its legends go back to Native American tribes and various people report sightings of it here and there. Many have gone hunting for the "creature", but are unable to catch it if it exists. Today, it continues on as a compelling enigma with sides pro and con about it. Related subjects are the Yeti, Orang Pendek, Yeren, Hibagon, Chuchunaa, and other forest manbeasts of lore. Today, this subject is regarded as pseudoscience with many practitioners.

Patterson-Gimlin Film Discussion

In the 1960s, two adventurers in the Pacific Northwest claimed to have filmed Bigfoot. They produced a questionable film showing what appears to be a humanoid creature running away into the woods. This film has since gone on to be world famous over whether it's a hoax or a real film of Bigfoot. For now, it exists in a world where it is routinely denounced as a hoax and where many scientists think it is genuine. If real, it is one of the great films of history. If a hoax, its one of the greatest hoaxes of history. For now, its controversy continues as to whether Bigfoot is real or not. Other controversies like it are the Myakka Ape, Yeti hand and skull, Orang Pendek, and Jacobs Creature.

Radio

In the 19th century, inventors would discover that a kind of radiation could be used to transmit signals and messages. It would lead to innovations like radio stations, radar, broadcasting, and other offshoots. Many inventors would delve into and pioneer all manner of invention from it. Today it manifests from walkie-talkies to aircraft radar. Its applications affect society in so many ways. Famous names in it are Marconi, Hertz, and many others.

Radio Astronomy

Inventors would try to use radio to study stars and astronomical objects. In this they would pioneer radio astronomy. They would build epic telescopes similar to giant dishes or dish antennas. Today its a vibrant area of astronomy known for pulsars, radio signals, and much more. Famous installations in this are: Arecibo, the VLA, Green Bank, Jodrell Bank, and others.

Physics "Mythology" Study

This refers to the study of dramas that made classic physics history (or physics mythology). These are dramas that have since become celebrated for epic discovery and epic moment in the history of physics. Famous dramas that have this quality are:

- Galileo's trial –Newton's apple –Trinity –Hiroshima and Nagasaki –Einstein's patent office –Einstein living at Princeton –Copernicus' deathbed –Gold foil experiment –Michelson-Morley experiment –Wolfgang Pauli's life –World War 2 events –Robert Goddard's rocket experiments –Moon Landing –Kitty Hawk –Marie Curie's life –Bikini –Young's experiment –Benjamin Franklin's kite experiment –Penzias and Wilson -Tesla v. Edison -planet discovery dramas

Planet Colonization Discussion

This refers to ways to colonize the Moon, Mars, and other bodies in the solar system. Topics here include terraforming (making a planet seem like Earth), the mass driver, moon mining, asteroid mining, domed cities, Mars farming, and so on. While a domain of science fiction, it is a vibrant domain of discussion speculating on whether people will live on astronomical objects like the Moon, Mars, other planets, asteroids, etc.

Planet Discovery

In astronomy, various epic stories would appear where planets or other astronomical objects would be discovered. These stories would fill books and define astronomy's history with characters and dramas where something was found. Topics to explore are:
- Uranus -Neptune -Pluto -2003 UB 313 (Eris) -Robert Brown's career -asteroid discovery -comet discovery -Planet X -Nemesis -exoplanets

Physics Fiction

Science fiction is a literature into imagining fantastic adventures blending science with fictional stories. A sub-domain of this concerns fantastic adventures in the domain of physics.

Ig Nobel Prize Discussion

The Ig Nobel Prize is a "facetious parody" on the Nobel Prize for Physics. It is a "prize" to commemorate "garbage physics" and ceremonies on it are staged every year. Students of physics can study this issue for physics that should not be done and is deemed worthless to the science of physics.

Hindenburg Discussion

The Hindenburg was a Nazi zeppelin (or blimp) of the 1930s. It was traveling to Lakehurst, New Jersey, USA in the 1930s. It was a giant blimp that used hydrogen (a dangerously flammable gas) to keep itself aloft in the sky. It famously flew across the Atlantic Ocean only to crash in flames in New Jersey. Its destruction was filmed and is today an epic account of air disaster. The destruction of the Hindenburg was thought due to a spark or act of terrorism, but no one seems to know what destroyed it. Today, its issue is controversial as it is still discussed.

THE GRAND SCIENCE OF PHYSICS

19th Century Physics

This is the study of physics as it was known in the 19th century. The 19th century is a famous era known as the Victorian Age. Here physics was discussed before relativity and quantum physics appeared in the 20th century. Famous issues in this time were the Michelson-Morley Experiment and Maxwell's Laws.

Spectroscopy

Its known white light comes in colors known as the spectrum (rainbow). Scientists would use this to study stars and determine what kinds of chemicals composed such things. A science would arise exploiting this and this is known as spectroscopy. It would grow into a vibrant sub-domain of physics with branches like:

- spectrography -mass spectrometry -neutron spectroscopy -spectrometry

Newton Archives Study

Isaac Newton left an archive of papers, writings, and other works. He is famous for writing the Principia, the Opticks, and other works of legend in physics. Today, he is regarded as a legendary life in physics for his "miracle" contributions to science. He would live in England in the 17th century and define history in the Scientific Revolution, laws of motion, calculus, law of gravity, and other contributions. He would become an epic character of history for his monumental life and myriad contributions.

Kinesiology

This is a study in fitness about the body's movements and mechanics.

Standard Model Study

The Standard Model is a collection of theories describing today's understanding of subatomic forces. It is a complex body of beliefs and is studied intensively in physics today. Subjects of study here are called QED, QCD, electroweak theory, and so on.

BRANCHES, SUB-DISCIPLINES, AND OFFSHOOTS OF PHYSICS

Physics Policy

Many governments operated a department of "science" or similar organization. In these organizations, officials decide what physics proposals are to be funded and what is to be rejected. Here physics becomes something called "Big Science", the blend of physics and government to finance great physics projects or science missions (a space program is an example). Many physicists find careers as officials of government in such organizations. Sometimes, a physicist will go on to become the President's Science Policy Advisor, an eminent position in physics itself. There exists an eminent body called the National Academy of Sciences. This is a group of high ranking scientists "elected" to serve on a board advising the government on science policy.

Oceanographic Physics

This is a blend of physics and oceanography (the science of the oceans). Topics here are discussions on marine technology like SCUBA, deep sea diving, marine constructions, and so on.

Dolphin Sonar Study

Dolphins are famous for having their own "natural sonar" system to navigate in the seas. This capability is studied by the US Navy and scientists.

Cetacean Intelligence

There is a speculation that cetaceans (whales, dolphins) have an intelligence that rivals Man, but this cannot be proven. Various thinkers would study these animals in hopes of discovering some human intelligence capacity to them. For now this is an obscure domain of study.

Golf Physics

This is the study of golf, the movement of golf balls, swings, and so on. Famous golfers of lore are Tiger Woods, Jack Nicklaus, and many others.

Space Exploration

This is the discussion on exploring the depths of space with spaceships and probes. It is practiced with such organizations like NASA, ESA, JAXA, and Roskosmos.

Contact Discussion

This is a discussion on what it would be like to "contact" alien intelligences.

Exobiology

This is a study into whether alien life exists and what it may be like. For now no alien life has been discovered and thus only speculation consists of this study. There is a search in science to find alien life and this is known as SETI (or the search for extraterrestrial intelligence).

Mars, Moon Exploration

This is the discussion on what it means to explore the Moon and Mars.

Electron Beam Lithography

This is the use of an "electron beam" to etch in metal and other substances.

Thermodynamics

Heat is an epic energy of Nature. It is in reality infrared radiation. It is used in such things as fire, internal heating, insulation, nuclear power, and in other domains. It has been studied for ages resulting in basic laws of heat and epic inventions. It today is a vibrant domain of physics with branches like:
- thermography -black hole thermodynamics -thermoacoustics -thermal engineering

'Save The Earth' Movement

Its thought the Earth is in terrible danger due to global warming,

BRANCHES, SUB-DISCIPLINES, AND OFFSHOOTS OF PHYSICS

the ozone hole, overpopulation, and the problems caused by Man. This refers to a movement to encourage public awareness and action to fight these problems.

Tacoma Narrows Bridge Study

The Tacoma Narrows Bridge was a famous bridge that was observed "waving" in a famous film. This bridge collapsed into a river. Since then, the study of this bridge has been a celebrated drama in the engineering of bridges.

Gymnastics Study

This is the study of gymnasts and their motion.

Nuclear Weapons Research and Design

This is the study of how to make other kinds of atom bombs, more explosive atom bombs, and their technology. This is studied in military nuclear laboratories.

Galactic Club Discussion

This is the discussion on what the nature of a "galactic club" is. It refers to a "federation" of alien societies as found in science fiction. The Star Wars movies, Star Trek, and other scifi films illustrate what one is.

Exoplanetology

This is the study of exoplanets or planets orbiting other stars.

Prosthetics

This is the study and design of artificial limbs like arms and legs.

Cybersecurity

This is the study of how to make computers "secure" from hackers.

Electric Company Discussion

This is the study of how electric companies operate.

THE GRAND SCIENCE OF PHYSICS

Laser Science

The laser is an light beam used for welding, detection, and in other uses. It was an epic discovery of physics that saw many kinds of invention. Today there are many kinds of laser known and it is studied in optics, the science of light.

Laser Chemistry

This is the use of lasers in chemistry.

Laser Music

This is the use of lasers to make laser light shows, laser harps, and so on.

Fusion Research

Fusion is a nuclear process to join elements like hydrogen to produce immense amounts of energy. It is the energy source of the Sun, stars, and the H-bomb. There is a worldwide effort to harness it for nuclear power, but so far its effort is too challenging and thus has not succeeded.

Robotics

This is a technology into artificial machines capable of work. Famous topics in this are:
- androids -worker robots -computers -toys -science fiction robots -robot films

Chromatography

This is a tool in chemistry to determine colors and chemicals.

Buoyancy Study

This is the study of how masses float on water like buoys and boats.

Volcanology

This is a domain that studies the nature of volcanoes, eruptions, and related issues. Volcanoes are epic displays where lava erupts, mountains

BRANCHES, SUB-DISCIPLINES, AND OFFSHOOTS OF PHYSICS

form, and cataclysms known as eruptions happen. They are studied in geophysics, geology, and elsewhere.

Io Volcanology

Io (I O or ee oh) is a moon of Jupiter known for very dramatic volcanoes. Its volcanoes are studied in planetary science.

Nuclear Science

This refers to a branches of physics that collectively study the atomic nucleus and its derivatives. It includes nuclear physics, particle physics, nuclear chemistry, etc. It would be invented from explorations into the atom and nucleus. Inventions in this are the nuclear reactor and atomic bomb.

Vocology

This is the study of talking, singing, vocals, saying words, and the voicebox.

Broadcasting

This is the study of broadcasting on the TV, radio, and other mediums.

Mountaineering

This is the study of how to climb mountains, hills, and so on.

Animal Motion Study

This is the study of how animals move.

Crash Physics

This is the study of car crashes, railroad derailments, impacts, and so on.

Microphonics

This is the study of the microphone (voice-sound device).

Cycling

The bicycle was an epic invention into riding a contraption for travel. A bicycle would go on to become an object of physics and is thus studied for its mechanics, physics, and other qualities. Bicycles would inspire domains like:

- bike repair -bike racing -bike business -motorcycles -bike travel -bike vacations -biker groups

Geyser Dynamics

Geysers are hot water disturbances found in Iceland, Yellowstone National Park, and elsewhere. There mechanics and nature are studied in this subject.

Recycling

This is the study and activity of taking garbage, disposed metal, disposed plastics, and so on and "recycling" them for later use.

Ship Construction

This is the study, design, and construction of boats, ships, and so on.

Electric Animal Study

This is the study of animals known to generate their own electric shocks. Topics here are electric eels, electric catfish, and so on.

Physics Game Theory

This is a subject where game designers make games using physics themes. Try and devise video game concepts using physics.

Video Game Physics

This is the discussion of how video game characters and effects behave.

Einstein Study

Albert Einstein was so phenomenal of physicist, he left the world many writings and papers. These writings have been archived and are open to students who become Einstein scholars. Einstein is a fascinating man and many biographies exist on him. Please read about Albert Einstein.

Physics Music Composition

This is the activity of composing songs with physics themes. Related issues are composing Star Trek music themes, Star Wars music themes, and so on.

Video Game Design

Video games are games played on TV sets and computers. Making them is a worldwide and fun business. However people who make a living out of making video games are required to know physics. Today VGD requires skilled programmers, artists, writers, and designers who have a good education in physics.

Classroom Physics

This is the discussion on methods to teach physics in a classroom.

Microgravity Study

This refers to conditions where gravity is severely reduced of its "field strength". Here astronauts experience weightlessness. Simulated gravity refers to rotating "toruses" to make a gravity effect.

Gravity Theory

Gravity has intrigued many thinkers for being a force that pulls down and causes orbits. Many would study gravity trying to control it, to cause 'antigravity', and to harness it for inventions. Its main theories are Newton's theory and general relativity. It is thought there is more to learn about gravity. There exist many kinds of gravity theory in physics lore and they can be studied.

THE GRAND SCIENCE OF PHYSICS

Industrial Physics

This refers to discussions on physics in paper mills, car plants, and so on.

Scientific Integrity Study

There is a problem in science of thinkers faking theories, equations, laws, findings, and data and passing them off as "scientific research". It is an issue that needs to be studied and fought.

Fiber Optics

The inventor Charles K. Kao is known to have passed a light beam through a "glass wire". It was found that information could be transmitted over glass wires in vast amounts. This began the issue of fiber optics and would change communication over this. Kao was awarded the 2009 Nobel prize for physics over this.

Witricity

This refers to ways of transmitting electricity through "wireless" means.

Star Travel Study

This refers to discussions on how to travel to the stars. Topics here include Orion, sleeper ships, generation ships, Bracewell probes, ramjets, and other futuristic starships. It is found in science fiction and has its thinkers in Carl Sagan (of COSMOS fame), Arthur C. Clarke, Isaac Asimov, and many others.

Hypercane Discussion

A hypercane is a speculated type of "super-hurricane" thought to be caused by giant asteroids impacting the Earth. Its thought the Great Red Spot on Jupiter is a hypercane and the creation of such a "superstorm" can destroy a planet. The nature of such storms is unknown and is largely speculation.

BRANCHES, SUB-DISCIPLINES, AND OFFSHOOTS OF PHYSICS

Alien Being Design

This refers to efforts to devise imaginary alien beings or think about the nature of alien beings. Topics here include alien races, the design of Star Wars characters, and much more.

Cryogenics, Cryonics, Cryophysics

This is the study of how to cool masses to extremely low temperatures.

2012 Millenarianism

The year 2012 is claimed in Mayan calendar tradition as a year of doom or portent of tragedy. It has caused a doomsday fad that was briefly popular in the years 2000 through 2012.

Orgonomy

Wilhelm Reich, a contemporary scientist in Einstein's time sought to claim that an unknown energy called Orgone existed. He claimed he could make devices from orgone and the study of orgone be called orgonomy. For the time being his claims do not stand up to scrutiny and he is now regarded as pseudoscience.

Mystery Energies

Scientists throughout the age would claim to discover kinds of energy. However later on their claims would be denounced as there would be no proof. Today physics lore knows many kinds of controversy into energies thought to exist but cannot be proven to however. Famous issues of lore are:
- N rays -chi -prana -mana -cosmic energy -Earth energy -orgone -odic force -crystal energy -psi energy -kundalini -vril -pyramid power -zero point energy -Unruh radiation -dark energy

Superconductivity

Many physicists would cool matter to near Absolute Zero and discover that matter behaves strangely here. Here electricity can flow where there is no resistance and many novelties arise. Scientists would explore this phenomena and discover such things as superfluids, supersolids, the Bose-Einstein condensate, and the bosenova. Today it is a vibrant domain of exploration in physics.

High Temperature Superconductivity

This is the study of efforts to make superconductors at room temperatures or near to it.

Ceramic Superconductivity

This is the discussion on making superconductors out of clay-like substances.

Experimental (Emerging) Technologies

This is the study of technologies currently being developed or may arise in the future. Topics here are robots, spintronics, metamaterials, fullerenes, holoTV, machine creativity, and so on.

Invisibility Study

This is the study of ways to cause objects to become invisible (or not seen).

Non-Rocket Space Launch Study

This is the study of ways to get into space without the use of rockets. Topics here are solar sails, space elevators, space cannons, supermountains, and so on.

Space Tourism

This is a discussion on how people may exploit space for tourist trips and vacations.

Private Spaceflight

Space travel is a complex and expensive activity of nations. It can only be done through the sponsorship of a space agency or for military ends. Despite its expense, private companies would explore to travel into space by means of private rockets and for tourist causes.

Nuclear Terrorism

There is a modern fear in society that perhaps terrorists would gain a nuclear bomb. They could potentially blowup a city or hold a nation hostage to outrageous demands. While such an act has not happened, the fear is so pervasive that many national governments have organized efforts to identify and destroy organizations that could sponsor an act of nuclear terrorism.

Nuclear Bomb Market

In the Soviet Union, there would be a vast arsenal of nuclear weapons. When this nation vanished, it was replaced by the nation of Russia. Russia would be plunged into turmoil where its thought they could not secure their Soviet nuclear bombs. Its thought nuclear bombs were stolen and are now being sold to paramilitary groups, terrorists, and suspect organizations. Its thought in the future, this could result in an act of nuclear terrorism.

World War Three

Its known that two costly world wars have occurred. Since 1945, the world has lived with a fear that perhaps another world war may occur. Its thought this war would be a nuclear war and that many nations would die in such a war. Its thought the consequences of such a war would cause a nuclear winter, the extinction of Man, and the end of the world. Many nations would build arsenals of nuclear bombs and over the years many episodes occurred where this war could ignite. As of 2012, nothing has happened, but its possible history may know the event called World War 3.

Artificial Intelligence

This refers to discussion of computers and machines able to "mimic" human intelligence.

Quasiparticle study

This is the study of atomic and subatomic entities called "quasiparticles". They are a technical discussion into topics like excitons, phonons, rotons, and so on.

Physics Creativity Research, Creativity Research

Physics is filled with many words like space, time, force, mass, work, power, electricity, magnetism, weight, and so on. Creativity methods are ways to derive new ideas from words. By applying creativity methods to physics concepts, new ideas in physics can be found. This would found the new field called 'physics creativity research'. It is a new subject in its infancy. Ideas here are to apply prefixes to physics words. Examples include 'super' + bomb = superbomb, and 'anti' + time = antitime.

Superbombs

A bomb is an explosive device used in war. However speculation would arise to derive fantastic kinds of bomb that can explode with forces greater than any type of military bomb. Fantasy superbombs so conceived are:

- supernova bomb -nova bomb -cosmic bomb -planet bomb -star bomb -infinity bomb -eternal bomb -mind bomb -god bomb

Physics Projects Inventing

Physics is filled with many concepts like space, time, mass, energy, and so on. Inventors have used creativity methods to "blend" words together. An example are the words, gravity and telescope. They are unrelated to each other as gravity is a force and a telescope is a device for seeing images. When "blended" together, the word, gravity telescope is "invented". Scientists then consider this word and plan a project to build a gravity telescope or a telescope able to sense for gravity waves. A new branch of astronomy called gravitational wave astronomy is an extension of this idea. Many new kinds of idea are possible here with much room to explore.

BRANCHES, SUB-DISCIPLINES, AND OFFSHOOTS OF PHYSICS

Nuclear Astrophysics

Astrophysics is the combination of physics with astronomy. Nuclear astrophysics is the study of subatomic particles and how they factor in astrophysics.

Particle Cosmology

There is a discussion in physics that if a subatomic particle called the neutrino has 'mass', then it can somehow cause a gravitational pull throughout the universe that can 'collapse' the universe in an 'end' known as the 'Big Crunch'. In physics, its thought the nature of subatomic particles can somehow affect the nature of the universe. This study is known as particle cosmology.

Telecommunications

This is the study of devices able to communicate over long distances. Examples are telephones, walkie talkies, TV, radio, cell phones, PDA's, GPS, and satellites.

World's Fair Discussion

The World's Fair is a celebration and exhibition event to display technology and world culture. Organizing one is a hectic experience and filled with detail.

Scientometry

This is the study of how to categorize and measure science qualities.

Scientology

This is an epic religion found in the USA founded by famous science fiction author, L Ron Hubbard.

Science and Physics Ethics

This is the discussion on the "moral" and "right way" of how to use science and physics knowledge and technology.

Space Law, Moon Law, Law of the Sea
This is the study of how to use space, the Moon, and the sea "legally".

Solar Distillation
This is the study of how to purify water by using solar energy.

Electrician Business
This is the discussion on how to organize and run an electrical business.

Physics Psychology
This is the study of how physicists think about physics.

Elevator Design
This is the study of how to design elevators and escalators.

Kinesthetics
This is the study of motion, its elegance, and its harmony.

Hoaxology
This is the study of how to perpetrate hoaxes in physics. Ideas are to claim to discover an energy, to invent a technology, and so on. Famous issues are the Moon Landing Hoax, N rays, cold fusion, free energy, Tesla devices, and so on.

Resonance Discussion
Resonance is the idea of a mass object having waves that oscillate back and forth in it. Resonance is an intensely studied issue in physics today.

Yarkovsky Effect Study
The Yarkovsky Effect is an obscure and very technical subject in physics, it is the subject of its own study. The reader is encouraged to explore this issue further.

BRANCHES, SUB-DISCIPLINES, AND OFFSHOOTS OF PHYSICS

X ray Research

X rays were discovered by Wilhelm Roentgen and he won the Nobel prize over this. X rays have since been used in a wide variety of subjects like crystallography, medicine, and so on.

X Wave Research

There is in physics the idea of the X Wave or "wave" that travels faster than the speed of light, called superluminal. Although never discovered, it is a subject of discussion and study.

Anemometry

This is the study of wind speed.

Horology

This is the study of clocks, watches, time keeping devices, sun dials, calendars, and other subjects.

Calendrics

This is the study of calendars.

Atomic Clock Technology

Atomic clocks are a technology using radioactive elements. It has lead to the invention of clocks known for performance and accuracy unknown in most other kinds of clock.

Radioactivity

In the early 20th century, thinkers like Marie Curie, Becquerel, and others would discover that mysterious radiations arise from kinds of element. They would study this and discover elements like radium, polonium, and much more. It time they would invent the word 'radioactivity' to describe elements that give off radiation. They would discover that elements may 'decay' or break down. Research into this would lead to many inventions, the hazard of radiation poisoning, radon, and much more.

Radio-Carbon Dating

Radio-carbon dating is a method of determining the age of something that is an offshoot of nuclear technology. It allows scientists to "know" the age of ancient wood, fire ashes, and anything known to contain the element carbon. It is an effective byproduct of nuclear physics.

Dendrochronology

This is a method to determine the age of a tree by the number of "tree rings".

Laser Cooling

This is the technology of using lasers to "cool" atoms to ultra-cold temperatures.

Greek Invention Study

This is the study of inventions from ancient Greece like Archimedes' screw.

Physics Organization Study

Many physicists have organized institutes, "societies", and groups dedicated to physics or some branch of it. Famous groups are the American Institute of Physics, the Royal Society, the Physical Society, the Max Planck Institute, the Kaiser Wilhelm Institute, and many others. Such groups maintain offices, elect officers, publish journals, and organize "events" devoted to physics.

Great Monuments Discussion

The ancients knew about such fantastic monuments in their time that they called them the Seven Wonders of the World. In this era, there exists monuments like theirs that impress the modern age. In this discussion, topics here discuss how to build such monuments, who builds them, and the effort required to construct them. Famous great monuments of the modern era are:

BRANCHES, SUB-DISCIPLINES, AND OFFSHOOTS OF PHYSICS

- Mt. Rushmore –Crazy Horse –the Capitol –the Sears Tower –the Statue of Liberty –the Great Wall of China –Big Ben –Christ of the Andes –Easter Island –the Tonga Trilithon –Amida Buddha –Polonnaruwa –the Forbidden City –the White House –St. Louis Arch –Superdome –Parthenon of Tennessee –Eiffel Tower –Sistine Chapel –Taj Mahal

Altimetry

This is the study of measuring altitude or "height" or "elevation".

Viscimetry

This is the study and measure of viscous fluids.

Giza Pyramidology

The Great Pyramid of Giza is a fantastic object from Ancient Egypt. It is a monument built of tons of stone in the shape of a pyramid. Nearby this monument sits other pyramids and the Great Sphinx. Various thinkers have studied this object and have "found" qualities about the pyramid that seem to suggest the builders had known of physics knowledge. Today, this subject is filled with controversy, outlandish claims of aliens and Atlanteans, and much more.

Ancient Egyptian Engineering

Ancient Egypt had "engineers" who were able to build pyramids, giant statues, obelisks, and great temples. They however did not know about the methods of "modern engineering". This has lead to much mystery and discussion as to how the Ancient Egyptians built such monuments as Karnak, the pyramids, Great Sphinx, and many other places. A famous thinker in this study is the ancient engineer called Imhotep.

Tonks-Ghirardeau Gas Study

A "Tonks-Ghirardeau gas" is a state of matter that's predicted to exist, but has never been "created". It is a challenge today to be able to make this "gas" and the efforts to do it.

Planetary Engineering

This is the idea, challenge, and technology of "shaping" planets according to human purposes. It is an area filled with science fiction dreams and grand visions. Examples are digging canals on Mars, building colonies on the Moon, colonizing asteroids, terraforming Venus, and so on.

Warp Drive Theory

This is a humorous and "fictional" subject devoted to the fake star drive technology called "warp drive". It is "discussed" in such television shows as Star Trek and explores ways to propel starships.

Wave Mechanics

This is the study of 'waves' in physics. Kinds of waves are water waves, sound waves, vibration waves, light waves, electromagnetic waves, and so on. Topics here include amplitude, phase, wave speed, frequency, period, crests and troughs, and so on.

Electrorheology

This is the study of fluid flow in electric fields.

Rheology

This is the study of fluid flow.

Particle Physics

In the early 20th century, physicists would discover that subatomic particles existed. They would discover such things like the electron, neutron, proton, photon, quark, antimatter, and much more. This would lead to the invention of particle accelerators and great research facilities into studying them. Today, particle physics is a major domain of physics studying all things subatomic with many findings, inventions, particles, and issues.

BRANCHES, SUB-DISCIPLINES, AND OFFSHOOTS OF PHYSICS

Particle Detector Technology

Particle detectors are devices that "detect" or "see" subatomic particles. Used in particle physics, some kinds are the bubble chamber, cloud chamber, and others.

Electronics

This is a subject into technologies harnessing electric currents. It would lead to inventions like TV, video games, toasters, ovens, radio, gadgets, and much more. Today it is found everywhere manifesting as all manner of appliances sold in stores. Repairing electronic devices is an epic industry as well.

Circuit Analysis

This is the study of electric circuits and efforts to make engines, motors, and so on. Topics here are Ohm's law, Kirchoff's laws, circuit diagrams, and so on.

Photonics

This is a technology into using photons or light particle for inventions.

Mayan, Aztec, Inca Engineering

These ancient Native American civilizations are known to have left great pyramids and other monuments. They existed in a time when no "modern engineering knowledge" existed and could not use cranes and forklifts. This is the discussion of the methods they used to build their monuments.

Megalith Study

Stonehenge is a fantastic monument known in England. Around Europe exists many large stone monuments like Avebury, Carnac, Castlerigg, and so on. How did ancient peoples construct these large monuments of stone without modern technology?

THE GRAND SCIENCE OF PHYSICS

Physics Library Science

Physics libraries are found in most any college, science institute, and other places. They are more than just collections of textbooks and it is a challenge to staff such a place.

Electrician Education

Various people are employed to fix telephone lines, install electric devices, and do various electrical work. Called electricians, they perform an essential function for businesses that need to "electrify" their constructions. Electricianry is taught at technical colleges and takes years to learn.

Wormhole Study

In physics, a "wormhole" is a "bridge" between two regions of spacetime through a black hole. Called Einstein-Rosen bridges, they are "envisioned" to be a kind of hyperspace used in science fiction. Today, they are a compelling issue and a TV show called Star Trek: Deep Space Nine discusses it.

Navier-Stokes Equations Study

The Navier-Stokes equations are a series of equations about incompressible fluids. They are difficult to discuss and technical to investigate, but efforts continue to find solutions to them.

Gyroscopy

This is the study of tops, gyroscopes, and spinning objects.

Thought Experiments

This is the idea of thinking about how a process goes in physics. Einstein used them often and today, they are a vital part of physics study.

Catastrophe Theory

This is the study and modeling of "catastrophes" like volcanic eruptions, tsunamis, quakes, etc.

BRANCHES, SUB-DISCIPLINES, AND OFFSHOOTS OF PHYSICS

Simple Machines Study

This is the study of simple machines like levers, inclined planes, screws, wedges, and pulleys.

Music Technology

This is the study of technology used in music. Topics here are tuning forks, radio, synthesizers, drums, instruments, guitars, amplifiers, laser light displays, turntables, disco technology, and so on.

Big One Prediction Study

In southern California, there is a growing fear that a devastating earthquake will strike in the 2010s-2020s time period. Efforts are underway to try to predict when it will strike and to take precautions to survive this "killer quake" when it hits.

Quark Engineering

This is an activity in particle physics of "constructing" particles by rearranging quarks inside.

Neutron Star Nuclear Physics

A neutron star is thought to be like a giant atomic nucleus. This refers to ideas where a neutron star is then thought to be an atom. To study a neutron star is in some sense a way to study the nucleus of an atom.

Atom/Solar System Analogy

It was thought for decades that an atom and the solar system were "similar" in model. Hence discussions of one could lead to insight into the other. This is a recreational activity in physics.

Blackbody Discussion

A blackbody is an "object" that is colored black, emits no light, and emits no heat. Its discussion was once widespread and influenced people like Planck to found quantum mechanics.

THE GRAND SCIENCE OF PHYSICS

Optical Biopsy

This is the removal of tissue from organisms. The tissue is then studied using optics.

Quantum Tunneling Study

This is the study of the quantum effect called quantum tunneling.

Raman Spectroscopy

V. Raman was a physicist of the 1940s who investigated Raman scattering. He won a Nobel prize over this and inspired this field.

Surface Physics

This is the study of interactions that happen on the surface of metals and other substances.

Statistical Mechanics

This is a branch of quantum mechanics dealing with the Ising and Potts models.

Pomeranchuk Cooling

This is the study of ways of cooling helium to temperatures of 0.3 Kelvin.

Regge Calculus

This is a technical issue related to advanced physics and relativity. It is very difficult and a branch of mathematics, it can be explored in advanced physics textbooks.

Creativity Machine Study, Machine Creativity

Today scientists are trying to make computers that can invent, create, compose music, calculate physics equations, and do so much more. Machines able to invent are called "creativity machines" and they are a promising new technology to explore. Its thought such a computer may lead to a "Genius Machine" able to be a "mechanical genius at anything".

BRANCHES, SUB-DISCIPLINES, AND OFFSHOOTS OF PHYSICS

Polarimetry

This is the study of photons as they move and are measured. It is used in astronomy intensively along with other subjects like Stromgrem interferometry and photogrammetry.

Polar Research

This is the activity of science in places like the Arctic and Antarctic.

Physical and Geometric Optics

These are branches of optics studying how light moves in lines and in the physical properties of light.

Pyrotechnics

This is the study of fireworks, explosives, fire, and related issues. It is taught to people who make fireworks displays. It can be a dangerous hobby as explosions can occur.

Nuclear Bomb Pyrotechnics

This is a futuristic speculation whereby nuclear bombs are used to cause fireworks displays. No known act of such inventing or display is known to have occurred.

Gravity Research

This is the study of ways to use "antigravity", cause it, and the nature of gravity overall. For the time being, no antigravity force is known to exist, but levitation methods do. There is a foundation dedicated to its study.

Engineering and Architecture

These subjects intertwine with each other and are concerned with the design and building of structures. Engineers build skyscrapers, bridges, ships, cars, roads, dams, amusement parks, and so on. Architecture is the study of actually designing structures in documents known as "blueprints". There are many branches to engineering and colleges dedicated to it.

Arcology

This subject is a blend of architecture with ecology pioneered by the thinker, Paolo Soleri. It would result in the epic construction, Arcosanti found in the US state of Arizona. Today it continues as a fad in science.

Electrical Engineering

This is a branch of engineering dedicated to electrical technology like capacitors, transistors, transformers, power lines, power plants, and anything electrical.

Chemistry

This is the science of chemicals. However, chemistry shares with physics knowledge of atoms, chemicals, weight, and so much more. To understand chemistry requires some education in physics.

Philosophy of Space, Time, and Spacetime

This is a branch of philosophy dealing with discussions on the nature of space, time, and spacetime. It tends to have its works and is taught in many colleges.

Foundations of Relativity, Quantum Mechanics, and Physics

This is the study, discussion, and investigation into the "foundations" or very basics of relativity, quantum mechanics, and physics.

Obscurity Study

This is the study of random features of physics and discussing and exploring them. Examples are the discussion of physical constants, equations, a bouncing ball, a flash of light, or other such thing. It is a random and "relaxing" way to think about physics and perhaps discover something new.

BRANCHES, SUB-DISCIPLINES, AND OFFSHOOTS OF PHYSICS

Magnetic Resonance Imaging, Nuclear Magnetic Resonance

This is the discussion of a technology used in medicine. It is extremely difficult and technical and discusses taking pictures of people's organs and "insides" to determine ailments. Topics here include X ray scans, PET scan, CAT scan, and so on.

Pre-Universe Physics

This is the discussion of what "physics" was like before the universe existed. It is mainly a recreational activity in physics and does not claim to have discovered anything as nothing is known. Topics here include nature of the void, the nature of God, pre-time, pre-space, pre-mass, pre-energy, and reality before reality.

Weightlessness Study

Astronauts know from being orbit that a gravity pull does not exist. This causes them to "float" in the air as they live aboard space stations and spacecraft. Weightlessness is a challenge to study as it has lead to the invention of a lot of technology. Astronauts must cope with water flowing out of glasses and reality behaving in ways different than on Earth. Weightlessness is simulated in water tanks and in a plane called the "Vomit Comet". It is a curious experience from living in space.

Automotive Mechanics

This is the study of the physics of cars, buses, jeeps, tanks, and other transportation machines. Its study is required for those who become car repairmen and engineers.

Biophysics, Biological Physics

This is a blend of biology (study of life) with physics. It models the motion of animals and is taught in biology courses.

Geophysics

This is the blend of geology (the study of the Earth) with physics. It studies such things as earthquakes (seismology), volcanoes (volcanology), alien planets, plate tectonics (plate movements), and much more. It has many branches and is taught in geology schools.

Agrophysics

This is the blend of agriculture (farming) with physics. It studies how crops grow, soil, weather, and climate. It is taught to people who become farmers.

Meteorology

This is the study of weather, clouds, storms, and so on. Phenomena here are hurricanes, tornadoes, blizzards, clouds, hail, wind, and so much more. It is taught to people who become weather forecasters.

Weather Station Discussion

This subject deals with the design and operation of weather stations.

Cloud Physics

This is the study of clouds and how they form.

Flat Earth Society Discussion

This is an obscure group that maintains the Earth is flat or a flat disc. Although considered nonsensical in science, it is a "leftover" from a time when people really believed that the Earth was flat and not round. Today, evidence gained from space probes, around the world sailors, and globe makers all attest that the Earth is round.

Climatology

This is the study of climates or "regions" of the world known for their weather. Climate regions are the Arctic, the tropics, deserts, forests, tundra, mountain regions, and so on. It is studied in connection with weather.

Sport Physics

This is the study of games like baseball, basketball, football, and

BRANCHES, SUB-DISCIPLINES, AND OFFSHOOTS OF PHYSICS

so on. Topics include pitching, dribbling, bat swing, and kicking. It is taught to coaches and athletes in many sports.

Robotics

This is the activity of building machines to do work (aka robots). Topics here are robots, androids, automatons, and industrial robots. Major thinkers are Minsky and Isaac Asimov, it is taught in major science colleges.

Humanoid Robotics

This is the activity of designing and building robots that mimic human beings.

Science Fiction

This is a literature dealing with aliens, space adventures, star empires, other worlds, ray guns, and so much more. It is futuristic and outlandish of its stories. Major works here are Star Trek, Star Wars, War of the Worlds, and so on. Major writers here are HG Wells, Jules Verne, Asimov, Clarke, and so on. This is a subject for those who like to "dream" fantastic adventures in space and in the future.

Jules Verne Discussion

J. Verne was an eminent science fiction writer of the 19th century. His works have since become classics and have been made into movies.

Fantasy

This is a literature dealing with wizards, sorcery, dragons, witches, King Arthur, and ages blending the Middle Ages with magic. Major works here are Narnia, Harry Potter, Lord of the Rings, and Dungeons and Dragons. It is for writers who like to dream exotic and magical adventures.

Tolkienesque Fantasy

This is the discussion on JRR Tolkien's classic fantasy novels, The Lord of the Rings.

THE GRAND SCIENCE OF PHYSICS

Star Trek Study

In the 1960s, a TV producer named Gene Roddenberry created the science fiction TV show called Star Trek. It lasted three seasons from 1966 to 1969 and then was cancelled. Afterwards, a fan movement grew to popularize it and demand that it be aired on TV again. Since then, a fad has grown to result in other Star Trek series, Star Trek movies, Star Trek toys, and other merchandise. It is now a popular culture fad rivaling any other. Characters and issues famous from these TV shows are:

- Captain James T. Kirk –The Starship Enterprise –Mr. Spock –Scotty –McCoy –Klingons –Vulcan philosophy –tribbles – Romulans –the Next Generation –Deep Space Nine –Voyager –Enterprise –Khan –Q the alien –Cardassians –Ferengi –the Dominion –Borg

Vulcan Philosophy

This is an offshoot of Star Trek that "discusses" the "philosophy" of the "planet Vulcan". Vulcans are a race of Trek aliens with "famous" members like Mr. Spock, Sarek, Surak, T'Pau, and T'Pring (Spock's wife). Vulcan philosophy discusses issues like logic, calmness, Zen attitudes, meaning, science, math, and related issues.

Klingon Philosophy

Klingons are another fictional race in Star Trek known for war, honor, barbarian ways, and much more. Klingon philosophy discusses honor, war, death, and loyalty.

Star Wars Study

In 1977, movie producer George Lucas created the science fiction film Star Wars. It created a fad in films, science fiction, and video games to follow after. Today, the series of movies of it have created a popular culture franchise that's been described as sensational. Characters and issues of these films are:

- The Force –Obi wan Kenobi –Luke Skywalker –Princess Leia –Han Solo –Darth Vader –Emperor Palpatine –Qui Gonn

Jinn –Chewbacca –Boba Fett –Jabba the Hutt –the Milennium Falcon spaceship –the Empire –the Rebellion –Death Star –Yoda

Dr. Who Study

In the 1960s in Britain, a TV show was created featuring a comic alien "Time Lord" known simply as the "Doctor" from planet Gallifrey. He travels in a spaceship with companions going on comic adventures battling villains and alien races trying to save the universe. At times, he will "die" and be "reincarnated" as another actor. Various actors named Tom Baker, Colin Baker, Peter Davisson, Jon Pertwee, and many others have played the Doctor. Today, Dr. Who is a cult fad in Britain among kids. Characters and issues of it are the following:
- the Master –TARDIS –Daleks –Cybermen –Kraag –companions –robots –Ace –evil –sonic screw driver –strange clothing –timespace travel –jokes –funny hats

Stargate Study

In 1995, there appeared a science fiction film called Stargate. It featured the discovery of a stargate or technology allowing hyperspace travel throughout the universe. A team of US soldiers and assistants travel through the stargate to an alien world ruled by the alien Ra. They have orders to blowup the stargate with an atomic bomb if they discovered anything of intelligence. Instead they become entangled in an intrigue to free a planet of people oppressed by Ra eventually destroying Ra with an atomic bomb. This movie becomes very popular inspiring a TV show named Stargate SG-1 and follow-ups Stargate: Atlantis and Stargate: Universe.

Nostradamus Discussion

In the 16th century, there lived a healer, sage, and mystic named Michel de Nostradame ('mishel day nahstradahm') or more famously as Nostradamus. He lived in a time when Europe was gripped by the

Inquisition, the Black Death, hysteria, death, and misery. He is known to have famously written a book called The Centuries. They supposedly contain prophecies of the future of the centuries to come. He wrote his book in cryptic statements to confuse readers and analysts to follow. Many readers of this book who are versed in history have gone on to say his "prophecies" seem hauntingly close to historical events like World War 2, JFK's assassination, the Moon Landing, and so on. Many TV shows have been made on Nostradamus and he continues as a popular culture fad even now. Today, modern scientists denounce Nostradamus as a hoaxer, but fascination for his writings continues on even now. Topics of discussion here are:

- Nostradamus seems to predict events of history over the last 400 years after his time.
- He is credited with predicting three tyrannical leaders that go by cryptic names similar to such historical personages as Napoleon or Hitler. His tyrannical leaders are given the name of "antichrist" for their "evil ways and deeds". Curiosities are the following:
- Napolon roi (Napoleon the king) –Hister (Hitler) –Mabus (unidentified villainous leader)

Please explore the issue of Nostradamus further as there are many books on him.

John F. Kennedy's Assassination

This refers to discussion on how President Kennedy was assassinated. Various discussions go that Oswald was not the real killer, that another gunman existed in a place called the "grassy knoll". There also exist films on this epic tragedy most notably the Zapruder film. Today, JFK's assassination is filled with intrigue over conspiracy theories, other gunmen, Soviet involvement, and so on.

Lincoln's Assassination

Abraham Lincoln was the president during the US Civil War. He was famously shot by John Wilkes Booth at Ford's theatre. Since then, this epic tragedy has been discussed over intrigues like co-conspirators, Confederate involvement, and so on.

St. Malachy Discussion

There lived an Irish saint of the Roman Catholic Church named St. Malachy. He is credited with writing a book of prophecy allegedly predicting the lineage of Roman Catholic Popes to live over a thousand year span. His writings were lost for ages inside a Vatican library (headquarters of the Roman Catholic Church in Rome). However his book was discovered, analyzed, and found to be "hauntingly similar" to the actual history of the Popes. Today, Malachy continues on as a popular culture fad, but he is denounced by modern scientists as a "lunatic prophet" from the Middle Ages.

Iceberg Flow Study

This is the study of icebergs as they break off from glaciers and drift in the sea.

Quicksand, Sinkhole Study

This is the study of quicksand and sinkholes.

Whirl Phenomena Study

This is the study of phenomena that show "vortex nature" or "whirl". Topics here are dust devils, tornadoes, hurricanes, whirlpools, waterspots, whirlwinds, and so on.

Geomagnetism Study

This is the study of Earth's geomagnetic poles, magnetic poles, and related issues.

Futurology

This is the study of the future or events to come. It studies trends, relations, projections, and future dramas to be. It intertwines with science fiction immensely.

Einstein Cross Study

There is in astronomy a finding called the Einstein Cross. Its discussion is technical and worth exploring.

Quantum Information Science

This is the study of qubits and ways to build a quantum computer.

Skylight Technology

This is the discussion on tools about skylights.

Fireplace Design

This is the study and design of fireplaces, chimneys, and firepits.

Beam Theory

This is the discussion on light beams and their nature.

Econophysics

This is a blend of economics (the study of money and finances) with physics. It discusses financial trends, markets, forex, and much more. It is taught to economists and bankers.

Physics Cartoon Character Invention

Father Time is a favorite character that embodies time, his creation is by some unknown author ages ago. There is a speculation that perhaps other characters can be conceived. Examples are Mother Space, Sister Energy, Brother Mass, and so on.

Bomb Shelter Design and Construction

In the 1950s, there was a fear that a nuclear war would break out between the USA and Soviet Union. This caused a fad to make bomb shelters across America. Even now many believe this war will happen and have taken steps to build bomb shelters.

First Objects Study

This is the study of the "first objects" that were created in the universe after the Big Bang.

Physics Cinema

This is the study, design, and making of movies, TV shows, and other productions discussing physics.

Ufology

People for ages would see things in the sky called 'unidentified flying objects' or UFOs. It would arouse many curious studies, fantastic issues, and outrageous stories of alleged aliens on planet Earth. It is studied as an obscurity in science known for speculations that perhaps aliens are here on Earth. So far no proof of anything alien would ever appear, but there are many epical sub-domains to study like:

- alien abductions -crop circles (cereology) -animal mutilations -contactees -UFO religion -men in black -UFO crashes -government UFO coverup -Majestic 12 -Area 51 -Roswell Incident -alien entities -ancient astronauts -UFO books -disappearances -UFO research -ice circles -Raelianism

Contactee Discussion

In the 1950s and 1960s, various people made outrageous claims that they had made contact with aliens or aliens chose them to communicate messages to Man. Many went on to say that they had met "alien people" with blond hair and traveled in "beamships" (alternative name for UFO). For a brief time, these people called "contactees" were a brief fad inspiring books and cults. However no proof of their claims has withstood scrutiny and they were denounced by modern scientists. Famous names in this stunt-fad are George Adamski, Van Tassel, and others. The reader is encouraged to explore this fad and sub-issue of the UFO phenomena.

UFO Religion

During the contactee fad, various people would go about beginning 'cults' saying aliens were here on Earth. These aliens were spreading religious messages, were contacting people, and that a 'new age' was coming. So far no proof of anything alien would ever appear, however movements known as UFO religions would arise allegedly claiming contact with an alien of some kind. Famous movements in this are:

- Raelianism -Urantia -Unarius

Geon Discussion

John Wheeler, a famous physicist and thinker in gravity proposed the idea of a particle called the "geon". It is a technical issue and has never been discovered. The reader is urged to explore this curiosity issue in physics lore.

Null Gravity Chemistry

This is a novel branch of chemistry that explores the nature of chemicals in the absence of a gravity field (weightlessness). Research into it can only be conducted in orbit and hence it is a costly kind of science. As of now, it is in its infancy as it awaits new explorers.

Astronomy

Over the ages, many thinkers would observe the stars, night time sky, and more. They would watch for comets, planets, and other things. They would in time found the science of astronomy. Since then astronomy would grow into an immense science where it is now so entwined with physics as to be a branch of physics. It is has many sub-branches like the following:

- gravitational astronomy -neutrino astronomy -radio astronomy -cosmic ray astronomy -galactic astronomy -solar astronomy -selenology (the Moon)
- cometography -asteroid science -planetary science -interferometry -astrometry -exoplanetology -black hole physics -cosmology

BRANCHES, SUB-DISCIPLINES, AND OFFSHOOTS OF PHYSICS

Astrology

Over the ages, various thinkers believed star positions could influence people's fortunes and future. This would lead to all manner of superstitious thing like fortune cookies, horoscopes, crystal ball gazing, geomancy, and so on. Astrology would be one of these movements, but science would denounce it for its superstitious tendencies. Astronomy would have its origin in astrology.

Pseudoscience, Pseudo-Physics

There are in science and physics continuing issues denounced as being unsound, not real, amateurish, and without basis in Nature. They are dismissed in science as issues that are not considered science. Despite this, these issues continue on in science at times studied and then denounced again. Famous issues of pseudoscience are:
- perpetual motion -N rays -orgone -mystery energies -the paranormal -magic

Gravitational Astronomy

Albert Einstein in his theory of general relativity predicted the existence of a new form of energy, gravity waves. It was not discovered until the 1970s, but it lead to a new branch of astronomy. Various "telescopes" of gravity have been built and it is now a flourishing science.

Planetary Science

This is a science that blends geology and physics. It studies the planets (called planetology) and other bodies. It is explored via space probes. Carl Sagan is a famous name in this science and there is an organization dedicated to advancing it called the Planetary Society.

Heliophysics

This is a science that studies the Sun, the parent star of Earth. Subject matter here are sunspots, corona, fusion, and much more.

Bosenova Study

Bose-Einstein condensates are known for a type of phenomena called the "bosenova" which is poorly understood.

Selenology

This is the study of the Moon. It studies moonquakes, geology, Moon rocks, and things about the Moon. It is named for Selene, goddess of the Moon.

Selenography

This is the activity of mapping the Moon's surface.

Heliography

This is the activity of mapping the Sun's surface.

Physics History

This is the study of past events of physics. The life of Einstein, Galileo, Copernicus, and the stories of physics are explored here.

Alternative Gravity Theory

This is the study of various kinds of theory of gravity. Many thinkers besides Einstein have invented their own theory of gravity. However, tests of their theories can be "proven" and it is a "dark alley" in physics research. Major names here are Brans-Dicke, Ni, Yilmaz, Watt-Misner, and many others.

Satellite Discussion

This is the study, design, and building of satellites. Kinds of satellite are weather, spy, telecommunications, cosmology, astrophysics, space telescopes, and so on.

Road Construction

This is the discussion on how to build roads.

BRANCHES, SUB-DISCIPLINES, AND OFFSHOOTS OF PHYSICS

Atomic Physics

This is the study of the atom, chemicals, and their physical properties.

Molecular Physics

This is the study of atomic "bunches" called molecules and their physical properties.

Chemical Physics

This is similar to atomic physics, but studies more complex chemicals.

Nuclear Physics

This is the study of the atomic nucleus, radioactivity, radiation, ways to build atom bombs, ways to build atomic reactors, and so on. It was born about 1945 and is practiced mainly in nuclear reactors.

Navigation Technology

This is the study of technology to travel. Examples are lighthouses, cars, GPS, sonar, radar, flashlights, walkie talkies, compass, astrolabe, sextant, and so on.

Consumer Electronics

This is the study of electronic gadgets sold in stores. Examples are telephones, cell phones, VCRs, computers, stereo, radio, ipod, Walkman, Blackberry, CDs, DVD, and so on.

Semiconductor Study

This is the study of a class of substances that "barely" conduct electricity. They are used in the making of electronic devices.

Electrochemistry

This is the application of electricity to chemicals and to study their properties.

THE GRAND SCIENCE OF PHYSICS

Thermochemistry

This is the application of heat to chemicals and to study their properties.

Galactic Dynamics

This is the study of the motion of galaxies.

Alchemy

In ancient times, thinkers would believe they could turn lead into gold. However this would be debunked in science later on. Today alchemy exists as a curious study into superstitions into chemistry.

Stellar Alchemy

This is the study of how stars can "make" elements by nuclear fusion. It is not real alchemy, but only uses the alchemy name.

Cosmology

This is the study of the universe as a whole. Topics here include the Big Bang, inflation, cosmic acceleration, dark matter, etc.

Eschatology

This is a discussion on how the universe will end.

Loop Quantum Cosmology

This is the study of cosmology with a branch of physics called loop quantum gravity.

Alternative Cosmology

This is the study of theories and ideas into cosmologies that claim the Big Bang did not happen or something is profoundly wrong about how scientists believe about the universe. Topics here are steady state theory, quantum cosmology, non-Big Bang theory, and so on.

BRANCHES, SUB-DISCIPLINES, AND OFFSHOOTS OF PHYSICS

TV Show 'The Big Bang Theory'

In the 2000s, a TV show would appear on college students who would live in an apartment. Here they would play games, video games, go to parties, have relationships, and have college buddies into science. This TV show would prove a sensation due to its comedy, guest stars, and other gimmicks.

Mechanics (Kinematics) (Statics and Dynamics)

This is the study of motion, structure, force, and falling objects. It is a required subject to engineering students. It is divided into two branches called statics (the study of motionless things) and dynamics (the study of changing things). It is a well developed branch of physics taught early on.

Neutrino Astronomy

Neutrinos ("little neutral one") are very small particles of mass. They are so small in fact that they are very difficult to "detect". Various scientists have built telescopes to find them that consist of vats of water buried deep below the ground. Today, it is a lively area of physics.

Appliance Repair

Electronic devices like TVs, microwave ovens, dishwashers, radioes, and so on break down. Many people are employed to fix broken machines. Shops dedicated to appliance repair exist in many communities.

Plumbing

This is the profession of employing people to work on pipes, clogs, drains, sewers, and so on. It studies how water flows due to gravity and pipes in general.

Spintronics

This is a new and innovative area of physics dealing with "electrical devices of spin". It is highly technical and can be explored further.

THE GRAND SCIENCE OF PHYSICS

Free Energy Research

Free energy is a controversial idea that the Earth has an "unseen" kind of energy that can be tapped for power. It is filled with hobbyists, investigators, and numerous "suspicious" claims. It can hold the promise of a new and abundant energy supply or be a fraud.

Aeronautics

This is the study of airplanes, helicopters, balloons, and other flying machine. It is for people who go on to be pilots and mechanics of airplanes.

Astronautics

This is the study of spacecraft, how to build them, their machinery, and so on. People here go to work for NASA, become astronauts, and so on.

Medical Physics Technology

This is the study of machines used in medicine. Examples are X ray machines, CAT scan, PET scan, heart machines, and so on.

Energy and Particle Astronomy

These are branches of astronomy where telescopes are built to "sense" in different kinds of energy. Examples are X rays, infrared, ultraviolet, gamma rays, cosmic rays, protons, neutrons, electrons, neutrinos, and so on.

Organized Self Criticality

This is a technical term for studying "breaking points". An example is a camel with lots of hay on its back. The camel is studied for where its breaking point is as to how much hay or straw it can carry. The "last straw" then is a piece of hay placed on the camel's back that causes the animal to collapse. Self-organized criticality is a word about how much stress something can take before it collapses or "snaps" into anger, helplessness, or inability.

BRANCHES, SUB-DISCIPLINES, AND OFFSHOOTS OF PHYSICS

Turbulence Study

Turbulence is when a plane encounters air currents that force it to fly awkwardly or when ships are forced to navigate in rough seas. It is the study of "chaotic" situations when a plane, ship, car, or device has to navigate.

Hoyle-Narlikar Cosmology

This is the study of beliefs in cosmology attributed to the scientists, Hoyle and Narlikar.

Parapsychology

There is in science a vast body of exotic and controversial subjects labelled under the words 'paranormal', the psychic, or psi. For ages people would believe the psychic was possible and it would be studied in science on occassion. However none of its claims would ever be proven real and thus it remains as an anomaly in science and physics overall. Famous issues of its study are:

- reincarnation -the near death experience -prophecy -precognition -levitation -mind over matter -psychokinesis -the out of body experience

Psielectronics

This is the idea that if the paranormal were a reality, then technology could be devised to "harness" it in some way.

Creationism Study

Creationism is an intense and controversial subject in science. Its discussions on the Bible vs. evolution, intelligent design, and so on have lead to many arguments.

Metal Detecting

Various thinkers realized you could build devices that could "detect" metal objects below the ground. This gave rise to metal detectors. They are used in treasure hunting and for detecting land mines. It is now a highly developed technology with many hobbyists. Famous brand names here are White's, Tesoro, OKM, Garrett's, etc. Specialities in this subject are gold mining, relic hunting, silver hunting, geocache, sunken ships, oil drilling, meteorite hunting, and so on.

Ham Radio

This is an offshoot of radio technology. It is an activity of hobbyists to use shortwave radios.

Plasma Research

This is the study of states of matter called plasma. A plasma is an ionized gas. Examples are lightning, fire, flames, and so on.

Military Technology

Physics has been applied to make weapons. Examples are atomic bombs, H-bombs, missiles, particle beam weapons, grenades, guns, mines, and other devices. It is studied at military colleges and national laboratories.

Archery

This is the study and use of the bow and arrow, its dynamics, and technology.

Police and Fire Science

These are areas where police and firefighters use technology based on physics. Stun guns, handcuffs, hoses, flame retardants, and other devices derive from this study.

Optics

This is the science of light and is a very ancient study. It had its origins in ancient times. It lead to such things as lenses, mirrors, telescopes, microscopes, and so on. Offshoots of this are used in medicine and became the subjects of optometry, ophthalmology, opticianry, and so on.

BRANCHES, SUB-DISCIPLINES, AND OFFSHOOTS OF PHYSICS

Optoelectronics

This is a subject combining electronics (electric technology) with optics. Products of this are optical computers (computers using light "circuitry), fiber optics (light "cables"), and so on.

Nanotechnology

This is a new and vibrant area of physics and technology with much possibility. It is the study of technology "tools" that are as small as atoms, cells, molecules, and so on. Inventors are hard at work making robots as small as cells and other futuristic things. It has many sub-branches and one is called femtotechnology, the technology of tools even smaller than this (the "nano" scale).

Atomtronics

This is the technology of making devices the size of atoms.

Animatronics

This is the design and construction of "robot puppets" used in amusement parks, movies, props, and other domains for entertainment purposes.

Mathematical (Computational) Physics

This is the application of math and geometry to physics. It has lead to math modeling, computer modeling, and much more. It tends to be very technical and requires years of study.

Quantum Chemistry

This is the application of quantum mechanics to chemistry.

Physics Education

This is the study of methods to teach physics.

Meteoritics

This is a study into meteorites (rocks from space) and attendant issues.

Photography

This is the study of cameras, films, film development, and picture taking. It is a lively domain filled with hobbyists, portrait studios, and so on.

Paper Science

This is the study of paper and the paper making process. It involves chemistry and physics.

Electron Microscopy

Louis de Broglie realized that matter (or electrons) has a wave and particle nature. This lead inventors to make "microscopes" using electrons instead of light. This has lead to novel kinds of photographs of very small objects. Electron microscopes are found in many physics laboratories worldwide. A Nobel prize was given over their invention as well as for another kind of microscope called the scanning tunneling microscope.

Electrostatics and Electrodynamics (Classical not Quantum)

This is the study of electric fields that change and do not change. It is studied in electricity courses.

Magnetostatics and Magnetodynamics

This is the study of magnetic fields that change and do not change.

Solar Cooking

This is the design and building of devices that concentrate sunlight to cause heat or fire to form. This is then used to start fires, cook meals or meat, or produce warming.

Arson Investigation

Arson is the criminal act of starting fires to burn forests, homes, and buildings. Pranksters or others engage in this type of crime. Police units act to investigate these acts and track how and who started "criminal fires".

BRANCHES, SUB-DISCIPLINES, AND OFFSHOOTS OF PHYSICS

Wildfire Fighting

Wildfires can be caused by arson, lightning strikes, or accidents. They are fires blown by wind and can burn vast areas and cause a lot of property damage. Wildfires are studied intensively and ways to fight them have been devised.

Survival Research

This is the study of the issue of reincarnation.

Coronography

This is the study of how to photograph the Sun's atmosphere, sunlight, and the corona.

Quantum Cryptography

There is in physics the idea of the quantum computer. It is an experimental device rapidly advancing in technology. A branch of this deals with ways to create secret codes to make computers "safe" from hackers. Called cryptography, it is an old subject dealing with how to make codes and keep secrets safe using quantum effects.

Field Theory

This is the study of fields (regions of space with an energy presence) and theories about them. It is a vibrant field with many offshoots like quantum field theory (or quantum fields).

Superstring Theory

This is the study of an entity called the "superstring". It is not known to exist and is thought to be an "entity of space" that vibrates. Branches of it are hoped to be able to one day produce the grand unified theory. It is today a vibrant area of research in physics with new developments.

M Theory

This is an offshoot of superstring theory and is known as one of physics' most hardest subjects. M is thought to mean anything from magic to mystery though it seems no one knows just what it means. It studies higher dimensional space and is involved in efforts to construct a grand unified theory. Its major supporter is Ed Witten, a thinker described as a "modern legend" in physics today.

Technology

This is the activity of inventing, studying, and using tools. Tools are devices that do work, perform tasks, and make chores easier. Tools come in kinds like the wheel, hand tools, gadgets, transportation devices, robots, computers, appliances, and so much more.

Satellite Technology

This is the study of devices used on satellites. This can include rockets, thrusters, solar panels, batteries, sensors, and other technology. It is a branch of space engineering overall.

Clarkian Science Fiction

This is the study of works of science fiction by Arthur C. Clarke.

Balloon Physics

This is the study of balloons. Topics here are balloon motion, helium and hydrogen buoyancy, materials, weather, and so on.

Ballistics

This is the study of guns, bullets, projectiles, cannons, and related issues. It is for anyone who has a fascination for guns and how they work.

Cartoon Physics

This is the study of how to make cartoon films, animation, and so on. Topics include coloring, animated motion, and so on and it is taught in film schools and college campuses.

BRANCHES, SUB-DISCIPLINES, AND OFFSHOOTS OF PHYSICS

Tired Light Cosmology

This is the idea that light as it wanders the universe is "pulled" on by gravity fields. This causes light to lose energy and hence become "weaker" or "tired". This has implications for how scientists understand the universe, it is a curious issue in cosmology.

Archeoastronomy (Ancient Astronomy)

This is the study of astronomy "heritages" of ancient societies like the Aztecs, Inca, Babylonians, Romans, Greeks, Chinese, Indian, and so on.

Magickology

This is the study of "magic tricks", illusions, and other performances as used by stage magicians, conjurers, and illusionists. Major names here are Houdini and David Copperfield.

Quantum Cosmology

This is the study of cosmology joined with quantum mechanics. It discusses how the universe may have born from a "quantum jump" and so on. Topics here include inflation, the Big Bang, and so on.

Quantum Electrodynamics (called QED)

This is a branch of physics called the "quantum theory of electric field changes". It is a discussion joining quantum mechanics and field theory and is a highly technical. Its main thinkers are Richard Feynman, Sin-Itiro Tomonaga, and Julian Schwinger who won the Nobel prize over inventing it. Another thinker named Freeman Dyson is also known to have made major contributions to its study.

Quantum Chromodynamics (called QCD)

This is a branch of physics called the "quantum theory of color force changes". It is a discussion on the strong nuclear force, quarks, nuclei, the "color" force (not light colors), and so on. It is a highly technical domain and many of its thinkers have won Nobel prizes over it. Famous names here are Gell-Mann, Zweig, and others.

Freeman Dyson Discussion

Freeman Dyson is a physicist known for making contributions to physics in QED. He is famous for inventing the Dyson Sphere (a science fiction sphere surrounding a star for energy production), writing books, and being a living legend in physics.

Extragalactic Astronomy

This is the study of galaxies, objects, voids, and other "structures" from outside the Milky Way Galaxy.

Undergraduate Research

This refers to research into physics performed by physics students while in college.

Graduate Research

This is the activity, study, and exploration of science projects performed by graduate students.

Science Fair Study

This is the study, organization, and involvement with science fairs or student competitions displaying science projects. Science fairs take place across the nation and are found in high schools and colleges seemingly everywhere.

Chaos Theory

This is a branch of physics dealing with chaos or disorder. It studies "systems" of masses that seem to be "insane" and attempts to "understand" them in some way. It is a highly complicated area.

Photogrammetry

This is a subject concerned with photographing stars, planets, and learning facts about them. It is taught as part of research into astronomy.

BRANCHES, SUB-DISCIPLINES, AND OFFSHOOTS OF PHYSICS

Observatory Design

Observatories are large buildings with domes that "house" giant telescopes. They are complicated structures that attempt to protect giant telescopes from the weather. There are many famous observatories like Keck, Mt. Wilson, Yerkes, the Lowell, and many others.

Godel's Incompleteness Theorems

Kurt Godel was a famous thinker, friend of Einstein, and a professor of mathematics. He is famous for writing what would be called the "incompleteness theorems" which influences science and logic. Please study their issues.

Telecommunications

This is the study of communications devices like telephones, walkie talkies, cell phones, and so on. It is a branch of radio research and allows people to "talk" over long distances.

Forensic Science

This is a subject that studies criminal cases for signs of evidence. Crime labs are where scientists study murder cases and the facts of criminal cases. Their findings can often decide a court case by what the evidence has to say.

Applied Physics

This is a "subject" more a "general name" for many sciences really. It is where physics thinking and findings are "applied" to technology and "real results" in the world.

Pure Physics

This is a "subject" that deals with physics thinking, investigations, composing theories, and working with equations. It is more for the "thinker" physicist than the "doer" physicist.

Interferometry

Literally translated as "the measure of interference", it is a collection of techniques in astronomy to photograph stars and study the sky. There are many sub-branches and each one is a highly technical discussion.

Atmospheric Physics

This is the study of the atmosphere, the movements of wind, behavior of clouds, and so on. It is intimately related to meteorology (the science of the weather).

Instrumentation Physics

This is the study of physics devices like centrifuges, Geiger counters, lab equipment, and so much more. It is filled with highly technical discussions.

Shock Physics

This is the study of shock waves, blast waves, extreme forces, impacts, and so on.

Photovoltaics

This is the study of solar cells and technology that can turn solar energy into "mechanical energy" for power. It was begun by Einstein and his work into the photoelectric effect.

Quantum Logic

This is the study of quantum effects and the "rules" that they follow. It is an obscure, difficult, and complex field with only a few practitioners.

Semiconductor Technology

This is the study of "microchips", silicon wafers, and circuits of computers. The integrated circuit is a topic of discussion here and there are many technology companies into it.

BRANCHES, SUB-DISCIPLINES, AND OFFSHOOTS OF PHYSICS

Theoretical Physics

This is the study and composing of physics theories by "pure thinking". It is explored in many colleges and topics here include M theory, superstring theory, relativity, and so on. Major thinkers here are Einstein, Hawking, Witten, and others.

Particle Astrophysics

This is a branch of physics that is a "blend" of particle physics and astrophysics. It discusses neutrino mass, photons, and other particles and their "influence" on stars and celestial objects.

Physics Philosophy

This is the study of the "meaning" of physics, that is it discusses the nature of space, time, spacetime, mass, energy, force, the universe, and so on. It is a union of philosophy with physics.

Tao Physics

This is the study of similarities of physics beliefs with Eastern Philosophy. Topics here include Zen, Taoism, Wu Li, chi, energy, formlessness, and so on. A major work here is called The Tao of Physics by author Fritjof Capra.

Quantum Reality Research

There is in physics discussions on the "meaning" of quantum mechanics. This has lead thinkers to speculate about "what is really happening" in quantum physics. It is a subject difficult to explain, but is filled with many controversies. Topics here include realism, neo-realism, nonlocality, Bell's theorem, oneness, implicate order, the Copenhagen Interpretation, and so on.

Quantum Gravity Research

There is a "dream" in physics to find a theory "uniting" quantum mechanics and general relativity. So far all attempts to "marry" these theories have been a futility and no one knows if it can be done.

Speculation suggests that a new understanding of gravity is needed or that a quantum theory of gravity must be invented. Today, it is a subject that could be a "dead end" of research and many offshoots are called "loop quantum gravity" and so on.

Condensate Study

In the early 20th century, two physicists Einstein and S. Bose worked on a new state of matter called a "condensate". A condensate is a state of matter where an atom seems to lose its "identity" and behaves like a "point" or "superatom". It is a highly technical discussion. Their research lain unknown for decades until other scientists produced a version of what would be called the Bose-Einstein Condensate. This inspired a "wave" of research leading to new condensates and Nobel prizes for its discoverers. Today, physicists are hard at work trying to make other kinds of condensate and study their nature.

Superfluid Study

Various physicists took to "cooling" matter to "supercold" temperatures as before. They found out that some matter (helium, a light gas) turns into a liquid or fluid like water. This fluid "flows" in ways water does not. Today, this liquid is called a "superfluid" and is a hot topic of research.

Bionics

This is an area of research combining electronics with biology. It involves the technology of artificial limbs, implants, and alterations to the human body to make it seem more like a machine than man. It has lead to the creation of cyborgs (cybernetic organisms), part man, part machine robots and other inventions. It is a branch of robotics. Famous TV shows exploring its fantasy are the Bionic Man, Bionic Woman, and Terminator movies.

Metaphysics

This is an ancient subject dating back to the time of the ancient Greeks and Aristotle. It is a discussion on fundamental concepts of physics interwoven with philosophy and "magical thinking". It has a developed literature and can be explored further.

Amateur Astronomy

This is the activity of astronomy as pursued by hobbyists "outside" of professional astronomy. Here people do their own stargazing, photographing stars and planets, hold star parties, and celebrate their fascination with "watching" the stars. They are known for finding comets, asteroids, new stars, and other celestial phenomena.

Celestial Mechanics

This is the study of how planets, comets, and stars move. Topics here are the laws of planetary motion, orbits, and so on.

Thoughtography Discussion

There is in study of the paranormal a discussion on how to make photographs by pure thought or paranormal means. It is today a subject embroiled in controversy as its riddled with hoaxers and outlandish claims. It is thought not real in modern physics.

Telepathy Discussion

There is in the study of the paranormal discussion on telepathy (or magical mind to mind communication). Its thought if telepathy exists it involves unknown powers of mind and brain. Claims of telepathy have not survived scrutiny and modern physics does not believe it exists. Today, people who claim telepathy are routinely disgraced as hoaxers and charlatans.

Spontaneous Human Combustion Discussion

There is in science discussion on people who are found burned to death by unknown means. Many thinkers claim these people are burned to death "spontaneously" or by some paranormal means. However no incident of SHC has been verified as paranormal in origin. Today, its thought victims of SHC merely died because they got caught in fires started by arsonists or by accident, hence no paranormal issue exists. It is a curious sub-issue in parapsychology.

Psychic Experiments by the US Government

There exists many rumors that US Government agencies like the CIA, Pentagon, NSA, Air Force, or military groups engage in psychic research. Its thought such agencies may try to explore "remote viewing", mind control, telepathy, contact with aliens, and so on. While many of these rumors are hoaxes, its thought some must be real going on. Today, its believed if the US government conducts psychic research, it is done by deluded government agents who have no understanding of physics and hence the unreality of these issues. It continues on as a scandalous intrigue in the US government.

Edgar Cayce Discussion

In the early 20th century there lived a prophet, healer, mystic, and counselor named Edgar Cayce ('kay see'). He is famous for giving "psychic readings" analyzing diseases people had, predicting Atlantis would be discovered, and so on. His caused a popular culture fad into his writings that continues on even now. His writings have been cataloged at a place called the Association for Research and Enlightenment. Today, Cayce is denounced and supported by many fans, although modern science does not believe he was a genuine psychic.

Crystallography

This is the activity of passing radiation (mainly X rays) through crystals to determine crystal structure. Many Nobel prizes have been awarded over this subject and it continues on as a vibrant branch of physics.

Nemesis Study

There is in astronomy outstanding mysteries about "unknown objects" lurking in the outer solar system. Speculation says that perhaps there is an unseen companion star to the Sun (called Nemesis), an unknown black hole companion (called Lucifer), unknown planets (planet Concord), and unknown dwarf planets or plutoids. In all, research is vibrant here as there is much to discover.

Black Hole Mechanics

Black holes are today a fascinating issue in astronomy. They are stars crushed by their own gravity to become something like "bottomless pits". There is a lot of effort to discover one, learn about their nature, and how they move. Stephen Hawking is famous for conceiving the "laws of black hole motion" or equations on how black holes move.

Black Hole Thermodynamics

This is the discussion of black holes and thermodynamics (the study of heat). It is found in astrophysics and is an obscure study primarily championed by Stephen Hawking.

Popular Science

This is the activity of writing about science and popularizing its subject matter. Products are magazines, websites, seminars, TV shows, movies, and so much more.

Solar Sail Technology

A solar sail is a "sheet" placed into space. It uses radiation and particles from the Sun as a mean to propel the sheet. It acts very much like a sail in using wind to move ships. It is considered a futuristic kind of space propulsion for moving spaceships. It has been tested once in a device known as Cosmos 1. Followups are the machine called Light Sail. It is a fantastic development in space technology.

Cosmic Ray Astronomy

Cosmic rays are extremely energetic "rays" from space. They have been studied via telescopes and are used to study stars and galaxies. A famous institute dedicated to this is the Pierre Auger Observatory.

Cytherology
This is the study of the planet Venus. Venus is known to be a hellish world with a runaway greenhouse effect. Many probes have been sent to map it.

Hermeology
This is the study of the planet Mercury. Mercury is a dead orb like the Moon known for craters, a magnetic field, and being near the Sun.

Areology
This is the study of the planet Mars. Mars is famous for its red color, being the possible home of life, having probes being sent to it, and for all the fantasies scifi thinkers have had with it.

Petrology, Petrophysics
This is the study of rocks, minerals, oil, and other "Earth" substances. Branches of this are gemology, mineralogy, petrophysics, and so on.

Speleology
This is the study of caves.

Zenology
This is the study of the planet Jupiter. Jupiter is by far the most massive planet. Galileo is credited with discovering four moons of it. It is known to have rings, giant storms, an intense magnetic field, and many moons.

Kronology
This is the study of the planet Saturn. Saturn is famous for its immense ring system.

Uranology
This is the study of the planet Uranus. Uranus was discovered by

BRANCHES, SUB-DISCIPLINES, AND OFFSHOOTS OF PHYSICS

Herschel in the 18th century and was intensely studied by the Voyager space probes.

Poseidology

This is the study of the planet Neptune. Neptune was found by LeVerrier, Galle, Adams, and others in the 19th century. It is colored blue with a great white spot on it.

Astronomical Art

This is the making of artwork depicting stars, planets, and "cosmic" scenes. It tends to be thrilling and involves painting and photography. It is something to explore.

Physics Art

This is the making of artwork depicting Einstein, Newton, Galileo, famous physicists, and "scenes" of physics in fascinating and artistic ways. It involves painting, collage, photography, and murals and can be a fascinating hobby.

Rocket Science

This is the study and building of rockets. Major names are Robert Goddard and Werner Von Braun (who lead the Nazi missile program).

Holography

This is the technology of holograms (or three dimensional photographs). They were invented by Dennis Gabor and have become a fad in science. Holograms are made by lasers and result in nifty photographs that seem to have a "depth". Today, holograms are found everywhere from credit cards to video games.

Radio Technology

This is the study of radioes and their technology. Examples of issues here are shortwave, ham radio, radio stations, broadcasting, and much more.

Bohmian Mechanics

David Bohm was a thinker and quantum physicist of the middle 20th century. He wrote about such issues like the Aharonov-Bohm effect, the implicate order, wholeness, and other issues related to quantum mechanics. He created a subject about the "mechanics" or "movement behaviors" of "quantum particles". It is a technical subject and can be explored further.

Gauss Study

Karl Friedrich Gauss lived in the 18th century in Germany. He is described as "genius" of science and influenced physics in many ways. His name is honored in physics in the electric unit, gauss. He is known to have mentored eminent mathematicians of his era, discovered Gauss' law, found the Gaussian distribution, and investigated the asteroid Ceres. In all, he lived an epic life of science and he is studied to this day.

Majestic-12 Study

In UFO research, there is an epic controversy called the Roswell affair. In 1947, supposedly a UFO crashed and was retrieved by the US military. It was covered up and "explained away" as a weather balloon. It is thought to have inspired a secret US government organization called Majestic-12 or MJ-12 which "hides" secrets of alien contact. Today, it is a controversial subject arousing many books and articles. Many physicists live second careers investigating this sensational issue.

Alien Abductions

There is a UFO issue where people claim to be abducted by aliens. These stories prove to be sensational yet unfounded. Studying them tends to be fascinating.

Antenna Theory

This is the study of radio antennas.

BRANCHES, SUB-DISCIPLINES, AND OFFSHOOTS OF PHYSICS

Brane Cosmology

There is in cosmology discussion on branes (or space-like sheets). They were derived from dropping the word fragment "mem-" from membrane (thus "brane"). Their discussion is interesting in that a collision of such entities may have caused the Big Bang. It is a highly technical area to explore.

Physical Science

This is a blend of physics, chemistry, geology, and other sciences taught to high school students. It is a subject to introduce students to the world of physics and other sciences.

Cometography

This is the study and photography of comets. Comets are thought to be asteroids with frozen ice and methane added. They originate in a faraway "place" called the Oort Cloud and "fly" into the inner solar system. They are famous for "glowing" and leaving a trail behind. In ancient times, the appearance of comets were thought to be omens of doom or disaster. Famous comets are Hale-Bopp, Ikeya-Seki, Shoemaker-Levy, Kahoutek, and Halley's comet.

Asteroid Study

Asteroids (or planetoids) were discovered by Piazzi in the 19th century. They are "primordial rocks" orbiting in between Mars and Jupiter. Many thousands have been discovered with many kinds of "asteroid belt" in the solar system. Issues connected with asteroids are the Death of the Dinosaurs, Meteor Crater, Tunguska event, craters and impacts, asteroid mining, asteroid colonization, asteroid Apophis, killer asteroids, and so on.

Anomaly Study

There are in physics many kinds of "mysterious" and controversial dramas and issues. Examples are N-rays, orgone, chi, pyramid power, new laws of motion, alternative gravity, antigravity, and so on. They are issues studied by many and yielding few results. It is an issue to explore.

Fringe Physics

This is the study and consideration into subjects that are on the "border" of reality, the known, the unknown, the respectable, and the outlandish in physics. These subjects are controversial and are known for intense debates and suspicious dramas. Examples are free energy, UFOs, the paranormal, miracles, Virgin Mary sightings, ghosts, exorcism, demonic possession, pyramid power, and so on.

Materialism

This is a discussion in philosophy into the nature of matter.

Aristotelian Philosophy

This is the discussion of views and beliefs held by the ancient Greek philosopher into the nature of the universe.

Substance Theory

This is the discussion in philosophy into the nature of substance.

Cosmogony

This is the discussion of any theory about the origin of the universe.

Philosophy Study

Philosophy is the study and discussion of fundamental ideas. It intertwines with physics immensely with many issues. Topics here include:
- Existence –logic –empiricim –logical positivism –phenomena –space –time –mass –energy –force –idealism –fate –Tao –Yin –Yang –structure –reason –ethics –worldview –causality –mind –essence –consciousness –free will –thing in itself –will –power –nature –evolution –being –nonbeing –creation –self –wheel of life –law –truth –good and evil –axiom –deduction –fallacy –awareness

BRANCHES, SUB-DISCIPLINES, AND OFFSHOOTS OF PHYSICS

Physics Camps

Many organizations run summer camps for kids teaching life skills, camping skills, and so on. A branch of this activity is the "physics camp", a summer camp for kids to learn physics. Activities here may be to study waves, motion, star gaze, learn about Einstein, and so on. It can be a fun place and something to expose to kids so they may learn about physics and decide if they want to study it.

Tomography

This is a technology associated with X-rays, medicine, and diagnosis of ailments practiced in medical colleges, clinics, and hospitals.

Star Gazing

This is the recreational activity of looking at the stars and the nighttime sky.

Naked Eye Astronomy

This is the practice of astronomy without use of telescopes, binoculars, cameras, or other equipment. It was practiced by ancient thinkers and continues on in star gazing and amateur astronomy.

Physics Heresy Discussion

This is the activity of discussing radical notions in physics. Topics here include how to overthrow relativity, how to overthrow quantum mechanics, how to overthrow any physics belief, and so on. It is mainly a recreational activity and is not considered a "real means" to start a "physics revolution".

Afterlife Discussion, Survival Research

This is the discussion on what comes after death. Is there a Heaven or Hell? It is mainly a recreational activity in philosophy classes.

Heaven and Hell Discussion

This is the discussion on the nature of realms or worlds that can be called Heaven or Hell. It arises out of religious studies and is a curious discussion at that filled with controversial notions and imaginative visions.

Killer Asteroid Study

Its known many asteroids pass by the Earth and some have hit the Earth in times past. Its thought asteroids caused the Death of the Dinosaurs and a famous place called Chicxulub is considered the "impact site". This subject investigates and monitors asteroids that perhaps could hit the Earth and wipe out civilization. Its feared an asteroid will hit the Earth eventually causing great disasters.

Ultimate Bomb

This is a discussion on using an atomic bomb to deflect a passing asteroid into a planet. It would have the affect of destroying a planet and causing an epic cataclysm. Its idea arises from the thinkers, Carl Sagan and Tim Voigt.

Asteroid Moon Creation

This is the discussion on how to harness asteroids that fly past the Earth and planets. It involves science fiction projects to make artificial moons out of them in orbit about the Earth and planets.

Weapons of Mass Destruction Study

This is the study and discussion of "weapons" known to be able to kill multitudes of people or "mass destroy". Examples of topics are atomic bombs, chemical weapons, biological weapons, dirty bombs, environmental terrorism, and so on. It is an issue discussed in studies of terrorism, crime, and anarchy and it's a major fear today that terrorists may acquire these weapons for "criminal acts". Please study the curious and potentially horrific issue called "American Hiroshima".

Mathematical Art

Physics relies on mathematics immensely for its ideas and practices.

BRANCHES, SUB-DISCIPLINES, AND OFFSHOOTS OF PHYSICS

Various artists have composed artwork exploring math, physics, and other science issues. Famous names here are MC Escher whose paintings are a kind of fad in science for his "challenging paintings".

Hypnotism, Mesmerism Study

Various people (magicians and illusionists) employ "methods" to "trick" people to recall memories, accept suggestions, and explore their "hidden selves". Hypnotism is used in subjects ranging from alien abductions to weight loss and is a curious study in itself.

Einstein Art

Albert Einstein has become such a "dominant" personality in physics known for his works on relativity and example in science. Various artists have taken to making portraits, statues, and other art work depicting him in "physics poses". Various actors make a living depicting Einstein and he has become an immense pop culture character all his own.

Astroseismology

This is the study of quakes in stars and planets (called "starquakes").

Women in Physics Study

Women are known to have made immense contributions to physics over the ages. People like Marie Curie, Marie Goeppart-Mayer, Lise Meitner, Mileva Maric, and others did incredible things and made epic stories. Today, women go on to become professors and thinkers even now helping to shape the science of physics.

Absorptiometry

This is the study and measure of absorption.

Theory of Everything

This refers to research, study, and efforts to find a grand unified theory or a theory that can explain all physical phenomena (or physical effects of the universe). A theory of everything (TOE) then is a theory that could "explain" all of physics in one grand vision.

9/11/01 Discussion

This refers to discussion on the 9/11/01 terrorist attacks in 2001. They were grim events that saw the destruction of the World Trade Center and the killing of nearly 3000 people. It is a grim issue connected to Osama Bin Laden and Al-Qaeda.

Carl Sagan Study

Carl Sagan was a prominent scientist of the late 20th century. He was a spokesman for SETI, involved with space exploration, founded the Planetary Society, worked on the Grand Tour, and made the COSMOS TV show, and wrote heavily. He has created for himself something of a fad and devoted fanbase to his works. Study and discussion on his life and works continue on in college and physics groups with more people learning about his legendary life in science.

Frank Drake Study

Frank Drake is a prominent scientist who founded the SETI movement, did Project Ozma, and leads the SETI movement. He invented the famous Drake Equation that discusses how many alien civilizations may exist in a galaxy.

Mayan Cosmology

This is the study of the ancient astronomy, worldview, and practices of "priest-astrologers-astronomers" of the ancient and extinct Mayan Empire. Other societies who are studied of their beliefs are the Persians, Sumerians, Babylonians, ancient Egyptians, ancient Chinese, Buddhists, Aztecs, Incas, Polynesians, and Native American tribes.

Europa and Exoplanet Study

There is a dream in science of finding evidence for alien life and civilizations. Though nothing has been found, it is a compelling "fantasy" that

BRANCHES, SUB-DISCIPLINES, AND OFFSHOOTS OF PHYSICS

has captured the imagination of many science fiction writers. It is believed in science that Jupiter's moon, Europa may harbor a "sea" of water. It is known that wherever liquid water is, its possible life may be found. Its thought Europa may harbor such life. Also, exoplanets (planets from outside the solar system) are thought to harbor life and civilizations. It is an activity in astronomy to search for such worlds and perhaps find alien life. Its also thought Mars may have hidden cavities with liquid water that may have fostered life. All this points to idea and dreams of continuing to search for alien life and the hope it will one day be discovered.

Space Engineering

This is a field of engineering concerned with building constructions in space. This has lead to such things as space stations, space probes, satellites, and space ships. Today, this field is changing and more is possible in this domain.

Metrology

This is the study of units, weights and measures, and other "measurement" qualities in science. Topics here include the metric system, the unit of mass, constants, and similar issues.

Vacuum Physics

The vacuum is just empty space. However, it is the subject of research into making technology to live in it, exploit it, and learn about its nature. Offshoots of it are vacuum cleaner technology, the nature of space, and so much more.

Vacuum Cleaner Technology

Many inventors noticed that a fan when whirring causes air to move. This creates a suction effect that can collect dust, leaves, particles, and so on. Many inventors have taken to making inventions called vacuum cleaners. They are found in households to clean rugs, carpets, and floors in minutes. They are used worldwide and found in most households. Famous inventors here are Oreck and others.

Logician Study

The logicians are a branch of Chinese philosophy known for "exotic statements" on the nature of reality. They are known for sayings like "fire is not hot". These are statements meant to get the thinker "thinking" about what he "understands" about Nature. It is a curious study to explore.

IMAX Theater Study

IMAX theaters are known for being visual experience that thrill and awe. They are such incredible theaters for their cost, enormity, and thrill to experience them. However, they are complex installations that are offshoots of optics, the science of light. Their discussion is highly technical, but the reader is encouraged to experience and visit one.

Television & VCR Repair

Televisions are complex gadgets that feature a "tube of gas" with energy inside to cause a picture on a "screen". VCRs are devices for playing "tapes" with movies and TV shows on it. These devices are used heavily in modern society and breakdown often. There is a lively business to repair broken machines like this and it intertwines with electronics.

Quantum Philosophy

This is a discussion on the "meaning" and interpretation of quantum mechanics, a highly complicated theory of physics. It is a tough subject practiced by quantum physicists.

Philosophy of Relativity

This is a discussion on the "meaning" and interpretation of the two relativity theories of Albert Einstein. It is a technical subject with intense discussions, it is practiced by physicists versed in relativity.

Numerical Relativity

This is the activity of finding solutions to relativity by using computers or number crunching.

BRANCHES, SUB-DISCIPLINES, AND OFFSHOOTS OF PHYSICS

Atomism
This is a discussion in philosophy and physics about atoms, small particles of mass. Issues here are the atomos, writings of Democritus, atomic thinkers, and much more.

LaGrangian and Hamiltonian Mechanics
This refers to the work of the thinkers LaGrange and Hamilton from years ago. They investigated Newton's equations and "reformulated" them in novel ways. Their work is studied today in mechanics.

Newtonian Cosmology
This is the study of the universe as it was known in Isaac Newton's time. It pays no reference to relativity, quantum theory, or modern research topics.

Colorimetry
This is the activity of measuring the colors of light.

Densitometry
This is the activity of measuring for density.

Lens Making
Lenses are pieces of glass. They come in varieties like convex and concave. They are known for focusing or "scattering" light in images. Lens making is the practice of "grinding" or shaping glass or other substances into lenses.

Glass Blowing
Glass is a substance known for letting light pass. Glass windows, glass cups, and glass objects all are examples of the value of glass. However there is an "art" that uses physics to "blow glass" or shape it into art objects and other things.

Stealth Technology

This is the technology to make airplanes "invisible" to radar. It is used by the military to design futuristic airplanes like the Stealth fighter and other aircraft.

Fustory

As history is the study of the past, then "fustory" is the study of the future or is a "history" of the future. This refers to events and dramas thought "inevitable" to happen in the future.

Geologic Fustory

This is the activity of predicting the future of planet Earth. It studies ancient Earth history and extrapolates the course of geologic history.

Illuminating Engineering

This is the study of lighting systems or the design of arrangements where lights "illuminate" a room. Topics here include flood lights, light houses, flash lights, forever flash lights, spot lights, etc.

Quantum Optics

This is the discussion of optics (the science of light) with quantum mechanics.

Classical Physics

This is a word referring to physics before relativity and quantum theory appeared. Its main topics are Newton's laws and electromagnetism.

Classical Mechanics

This is the study of Newton's laws of motion and ideas connected to them.

Orbital Mechanics

This is the study of objects and their motion as they orbit the Earth.

Space Plasma Physics

This is the study of plasmas (ionized gases) in space.

Gauge Theory

This is the study of physics topics like quantum field theory, invariance, and other issues. It is described as a difficult subject and virtually all textbooks on it are "tough reads".

Free Body Diagram Analysis

Free body diagrams are "pictures" of blocks and objects with "force arrows" being drawn on them. They are studied in physics to analyze what kinds of forces act on it.

Projectile Mechanics

Projectiles are objects like bullets, cannonballs, discus, and other "launched" objects. They are studied for their motion. Topics here include fall rate, acceleration due to gravity, and motion equations.

Thermoacoustics

This is the study of sound-heat devices. That is it studies machines that can turn heat into sound or sound into heat. It is a highly technical domain and an obscure branch of acoustics, the study of sound.

Space Station Engineering

Space stations are a modern wonder for being "outposts" in space for people. They are complex machines and not many were ever built. The ones known to history are Skylab, Almaz, Salyut, Mir, and the International Space Station. Skylab is known to have crashed back to Earth in the 1970s and the Soviet space station Mir was crashed as well. The International Space Station is a multi-national effort to build a space station for many nations and a sign of cooperation among nations.

Hydro Engineering

This is a branch of physics dealing with building such constructions as dams, bridges, sluice gates, "locks" and canals, and so on. Projects of this science are enormous and costly. Famous installations are the Hoover Dam, Aswan high dam, Panama Canal, Suez Canal, and so on.

Visionary Study

Physics and science overall are filled with people who are "visionaries" or "professional dreamers" whose job is to invent new gadgets, new science projects, ideas for space exploration, and so on. A famous example is Gerard K. O'Neill, a visionary known to have invented the O'Neill space colony, a wheel-like space station of fantasy. Science fiction authors are noted visionaries.

Fantasy Physics

This is a subject whereby "dreamers" conceive of all kinds of "fantasy" or fake physics, made up phenomena, and imaginary forces and particles. Example of fake physics are hyperspace, time tunnels, infinite speed, perpetual motion, the Death Star, the God particle, and so on. It is a subject where physics becomes science fiction and anything is possible.

Physics Computer Research

Computers are powerful machines able to calculate and process data. Many scientists have built computers able to solve physics equations, model physics phenomena like weather, and do other things. It is a vibrant area of research.

Mind Philosophy

This is a branch of philosophy that discusses the nature of thought, mind, concepts, thinkers, and knowledge. It discusses an idea called the "universal mind" or a "mind of the universe" that can think and make the universe. It is an interesting subject with many books. It intertwines with physics with many discussions.

Scientific Skepticism

This is a subject whereby scientists "professionally doubt" or "skoff" at claims of the paranormal, Bigfoot, outlandish claims of science, UFOs, N rays, perpetual motion machines, and so on. Various scientists attempt to act as "authorities" of science to lecture about the reality of issues. They assume the mantle of "expert" or "reasonable person" and test claims into speculative physics and science claims. Famous figures in this obscure movement are Carl Sagan and a group called the Committee for the Scientific Investigation into Claims of the Paranormal (CSICOP).

Perpetual Motion Study

Perpetual motion is the idea of motion that never ends. It discusses the making of "machines" that once are set in motion will be in motion forever. Today, it is believed perpetual motion cannot exist. However, inventors for centuries have tried to make such machines and get patents for their "creations". Many patent offices routinely "reject" any applications to patent a perpetual motion machine. Today, perpetual motion is a controversial subject and is in disrepute.

Esoteric Cosmology

This is the study of "offbeat" and unusual views of the universe. Topics here include Gnosticism, Urantia, Tantra, Theosophy, Max Theon beliefs, the Fourth Way, PaGaian cosmology, and Rosicrucianism. These beliefs blend the occult with cosmology and are considered very controversial issues to study and discuss.

Quantum Suicide Study

Schrodinger's Cat is a thought experiment about a cat being both alive and dead in quantum physics. Quantum suicide is an extension into the discussion of Schrodinger's Cat.

Quantum Mysticism

This is a subject that explores quantum mechanics, consciousness,

Eastern philosophy, Taoism, and other subjects. It is a controversial subject with much discussion.

Atomium Study

Atomium is an idea of an "atom" whereby a nucleus has been replaced with another kind of particle. An example is called "positronium" whereby a positron (antimatter electron) and an electron are in orbit about each other as an "atom". Other ideas are to make "atoms" with "nuclei" of particles like quarks, photons, neutrinos, pions, muons, and so on. It is today a curiosity in physics.

Antimatter Chemistry

The Periodic Table of the Elements is a chart of the "known" fundamental atoms of Nature. They go from Hydrogen (the lightest element) to Meitnerium (the heaviest element). There is a practice in physics to make atoms made of antimatter. At present an atom of "anti-hydrogen" has been made and it is possible that an "anti-Table of Elements" can be composed. It is a curiosity in physics.

Metallurgy

This is an ancient science that studies metals. Examples are precious metals, gold, silver, iron, steel, bronze, brass, electrum (a blend of gold and silver), alloys (mixtures of metals), and so on. It discusses such issues as plating, malleability (hammering), mining, and smelting.

Kirlian Photography

This is a branch of photography dealing with "images of auras". It is a controversial domain that seems to involve the occult and paranormal. The reader is encouraged to explore it further.

Wildlife Photography

This is the activity of trying to photograph and film wild animals.

BRANCHES, SUB-DISCIPLINES, AND OFFSHOOTS OF PHYSICS

Coral Castle Study

In the early 20th century in Florida, a "hermit" from Latvia named Edward Leeskalnin built an elaborate rock sculpture called Coral Castle. He is thought to have carved blocks of coral weighing many tons to make his constructions, however no one ever observed him actually making his "place". He lived a life as a recluse, loner, and man troubled with many problems until his death years later. Today, his construction survives as a tourist attraction called Coral Castle. The reader is urged to explore this story.

Hoax Photography

This is an activity where photographers make 'photos' depicting such 'things' as UFOs, ghosts, Bigfoot, the Loch Ness Monster, aliens, and so on. It is a study in trickery and trying to stage hoaxes. Topics of issue here are the Patterson-Gimlin film, UFO photographs, and Bigfoot photographs.

Nuclear Alchemy

Alchemy is an ancient superstitious practice of trying to change lead into gold. Although alchemy has never been known to be a "real science", it did give birth to chemistry (the science of chemicals). Today, particle physicists have used particle accelerators to change an atom of lead into gold. In some sense, alchemy has been achieved and it survives as a curiosity.

Big Bang Alchemy

This is the study of how the Big Bang could have made chemical elements.

Element Creation Study

There is a practice in physics of trying to make newer elements. An idea is to use a particle accelerator to "bombard" heavy atoms with protons to make newer elements. This process has lead to the discovery of new elements that have been named Seaborgium, Meitnerium, and so on.

Magnetic Levitation Study

Magnetic levitation (called "maglev") is a technology using magnets to cause objects to hover in the air. It is a "magical effect" that allows for such things as maglev trains and other technology. Today, maglev trains are used by many countries and this issue can be explored further.

Polygraphy

This is the practice of inventing "lie detector machines" or polygraphs. Polygraphs are used by police to question witnesses and criminals and is a controversial subject.

Thermionics

This is the study of heat technology.

Low Temperature Physics

This is a branch of physics that studies mass cooled to temperatures near Absolute Zero, the coldest possible temperature.

Astrogeology

This is the study of the "geology" of planets, stars, asteroids, and other celestial objects. Topics here are tectonic plates, volcanoes, trenches, plains, planet interiors, and so on.

Cereology

This is the study of the mystery of "crop circles".

Pendulum Study

Pendulums are strings with a weight (called a bob) at the end and are swung back and forth. They are curiosities in physics and have been studied for ages. Pendulums are used in hypnosis, clocks, and other devices. A thinker named Leon Foucault invented the famed Foucault pendulum and determined the Earth was spinning. Topics here are yo-yos, metronomes, and so on.

BRANCHES, SUB-DISCIPLINES, AND OFFSHOOTS OF PHYSICS

Physics Television Production

This is the study and activity of making TV shows teaching physics. Examples are such productions called Cosmos, Stephen Hawking's Universe, and so on.

Concept Car Design

This is the study and design of "designer or dream" cars using the latest technology. Topics here include solar cars, Lamborghini, Porsche, electric cars, pedal cars, and so on. Branches of this are to design concept airplanes, boats, balloons, bicycles, and other devices.

Riemannian Geometry

Bernhard Riemann was a mathematician of the 19th century. He studied "geometry" of surfaces that curved or were distorted. His work became famous in general relativity and was used by Einstein. It is a subject discussed in connection with a subject called Non-Euclidean Geometry and is a technical issue.

Fourier Study

Jean Baptiste Joseph Fourier was a leading thinker in waves, light, and heat in centuries past. His life and work have become something of a legend in physics and he is studied to this day. He is known for his work in waves and derivatives.

Supersymmetry

This is a research fad subject in advanced physics. It teaches subatomic particles have "counterpart particles" called superpartners. These particles are named by adding the prefix s- to them. For example, subatomic particles like electrons, protons, and neutrons have their "superpartners" named selectrons, sprotons, and sneutrons. This subject discusses many difficult issues like neutralinos, MSSM, supergravity, dark matter, vacuum energy, extra dimensions, and so on.

Astrometry

This is the study of how to locate stars in the sky.

Ideal Gas Study

An ideal gas is the notion of a "perfect gas" and this issue has been heavily discussed.

Periodic Table of The Elements (Chemistry)

There is in chemistry and physics a chart for "grouping" elements or kinds of atoms. Discovered by Dmitri Mendeleev of Russia, it is a chart vital to any education in physics and chemistry. It discusses elements or kinds of atom. Each atom has an amount of protons (atomic number) in its nucleus that defines what kind of element it is. Hydrogen (the lightest element) has one proton and occupies the "first box" on this chart. Helium (the next element after hydrogen) has two protons and occupies the second box on the chart. In all there are over 100 elements known and more are being discovered. A box on this chart lists the "symbol" for the element (letters) and discusses qualities important about each element. Elements have been discovered accidentally over the ages and thus they have all kinds of names like gold, silver, oxygen, neon, francium, and nobelium. New elements when they are found are added the suffix –ium and are named by their discoverers. Elements and all the things that can be done with them are what chemistry is about and learning this chart is important for any advancement in science.

Theory of Impetus

This refers to a body of ideas and discussions dating from ancient Greece. Major thinkers in this are Hipparchus, Philoponus, Jean Buridan, and Averroes. It is a discussion on force, resistance, and how projectiles (like cannonballs) move. It eventually lead to such ideas as inertia and momentum. Later on it "evolved" into the study of mechanics.

Circular Motion Study

This is the study and discussion of "spinning masses". It discusses ideas like centrifugal force, Coriolis force, fictitious force, spin, tops, gyroscopes, and so on. A centrifuge, merry go-round, and satellites in orbit illustrate examples of circular motion.

Tokamak Study

In Russia, various thinkers worked to build a "fusion reactor" using lasers and plasmas. This lead to the building of a machine called a "tokamak". Major thinkers here are Tamm, Sakharov, and others. It is today considered a promising technology to harness fusion and examples of such machines are found in Italy, Germany, Japan, the USA, and other countries.

Occult Movements Study

The occult or issues dealing with magic, spells, and superstitions has had a reputation of disrepute in science. It is not considered "real" or "practical", but nevertheless it has been heavily discussed in physics for its issues of magic, spirit, magical objects, scrying, and other issues. Magic has never been known to be real in science. However, ancient thinkers thought it was real and it intertwined with practices that would lead to physics. Today, the occult survives in primitive tribes, in voodoo, in astrology, and other vestiges of ancient times and ways.

Planck Scale Study

The Planck Scale is a word that refers to energies of $1.22 * 10^{28}$ eV and conditions from the very early universe. It is named for Max Planck of quantum theory fame. It is a "condition" where all four fundamental forces of Nature are "of the same strength". It is a technical discussion and one heavily researched in science today for "new physics" to explore.

Sub-Planck Physics

This refers to "physics" that is beyond or smaller than the Planck Scale. It studies "conditions" from the very early universe where "ordinary notions" of space and time do not seem to apply. It is today a speculative study filled with many mysteries.

Space Advocacy

This is the activity of supporting travel and projects in space. It involves testifying before Congress to finance space projects, petitioning NASA for space projects, and so on.

Space Health

This is the study of the "health" of astronauts as they live in space.

High Pressure Physics

This is the study of matter under enormous pressures.

High Temperature Physics

This is the study of matter of temperatures of hundreds to thousands of degrees.

Einstein-Cartan Theory

In the early 20th century, Albert Einstein and Elie Cartan collaborated on a theory later to be called Einstein-Cartan theory. It is today a topic in advanced physics and has many offshoots.

Luminescence Study

This is the study of kinds of "light emitting" phenomena. Topics include sonoluminescense (light by sound), bioluminescence (light in life, glowworms, fireflies), chemiluminescence (light by chemicals), and other issues.

Culinary Technology

This is the study of technology used in kitchens, restaurants, and so on. Examples are knives, stoves, grills, ovens, dishwashing technology, and so on.

BRANCHES, SUB-DISCIPLINES, AND OFFSHOOTS OF PHYSICS

Crush Technology

This is the study of mass under high pressure. Topics here include car crushers, pressure chambers, can crushers, and so on.

HVAC Technology

This refers to heating, vacuum, and air conditioning technology. People trained in this are called HVAC technicians and are usually repairmen paid to fix broken appliances.

Cable Television Installation

This is the activity of connecting homes to cable television providers.

Chess-Universe Study

Chess is a boardgame known for "checkmating the king" and is played by millions of people. This is an activity of comparing chess to the universe and physics. It is a "mental recreation" with many curious ideas being produced.

Universe Models

This is the activity of comparing the universe to such things as: the board game of chess, a loaf of bread, a trampoline, a television set, a blanket, and other ordinary objects.

Neon Technology

Neon is a noble gas known for glowing when "excited" by electricity. This has lead to neon lamps, neon street signs, advertisement displays, and so on.

Noble Gas Chemistry

Noble gases are chemicals ranging from Helium to Radon. They are known as "inert" chemicals for not "reacting" with other chemicals. Ways have been found to "coax" them to interacting with other chemicals and this lead to a new branch of chemistry.

Bomb Physics

Bombs are explosive weapons known for destroying things chaotically. Many scientists have programmed computers to "model" how bombs behave and how to build better bombs. This is pursued at military physics installations.

Strobe Technology

The strobe lamp is a "light" that spins around in a way to make motion seem easy to study. Its novel light effects make it fascinating and it used in casinos, dance halls, and elsewhere.

Extreme Sports

Sports are a term for such activities as baseball, football, and basketball. However, there are activities that "push the barrier" of sporting activity into a domain called "extreme sports". Examples of this are skydiving, zorbing, marathon running, iron man competitions, snowboarding, extreme skateboarding, surfing, bunjee jumping, and so on. These activities require technology that engineers have had to invent.

Bubble Study

Kids know about making bubbles with soapy water. Bubbles are interesting phenomena discussing tension, surface area, and so on and are studied in physics.

String Study

This is the study of wave phenomena of strings, ropes, twine, yarn, and so on. It is not related to superstring theory which is advanced physics.

Bicycle Mechanics

This is the study and design of bicycles, tricycles, and related devices.

Metallography
This is the study of metals and their shapes and issues.

Theoretical Physics
This is a domain to study advanced theories in physics. Topics here include investigating cosmology, trying to find the grand unified theory, composing a theory of quantum gravity, and more. It has many sub-domains like:
- M theory -supersymmetry -supergravity -twistor theory -Peccei Quinn theory -quantum gravity -loop quantum gravity
- An Exceptionally Simple Theory of Everything -theory of everything -string theory -superstring theory -Kaluza-Klein theory

Adaptive Optics
This is the activity of altering a telescope's lens to increase or decrease its ability to make images of stars and celestial objects.

Comparative Planetology
This is the comparing of planets for similarities and differences.

Core Accretion Theory
This is the belief that gas giant planets grew immense by their own gravity attracting mass to their cores.

Dynamo Theory
This is the idea that iron inside planet cores moves and causes a magnetic field.

Xerography (Xerox)
This is the study of copying machines and their technology.

Aeronomy
This is the study of aerosols, air flow, and related subjects.

Kaluza-Klein Theory

Theodor Kaluza is known for adding a "fifth dimension" to space in relativity and his work was improved by Klein. This created a research fad into their work called KK theory.

Theoretical Nuclear Physics

This is a branch of nuclear physics dealing with "thought" and models on what happens inside nuclei, atoms, nuclear reactors, and so on.

Resonance Theory

This is the study of resonance.

Audiology

This is the study of hearing, the ear, and how sound affects the ear.

Electroplating

This is the study of "coating" metals in layers of gold, silver, or some other metal.

Nanoengineering

This is the study of "engineering" at incredibly small distances.

Zodiac Study

The zodiac is a collection of "constellations" or random star patterns dubbed as a form or character. Examples of constellations are Pisces, Gemini, Aquarius, Libra, Taurus, and so on. In a subject called astrology the zodiac is used to "cast horoscopes" or superstitious advice on luck. Although the zodiac has no practical reality, it is used extensively in astrology and for centuries. Today, its studied as a curiosity in physics more for its "history" than for giving advice on anything.

Microscopy

This is the use of inventions called microscopes (or telescopes of

BRANCHES, SUB-DISCIPLINES, AND OFFSHOOTS OF PHYSICS

the small). These are devices used to look at cells, bacteria, grains of sand, and other "small things". Inventors have made all kinds of microscope including one called the scanning tunneling microscope which won a Nobel prize.

Maxwellian Relativity

This is the discussion of "relativity" in James Clerk Maxwell's time, well before any of the "real relativity" theories ever appeared.

Galilean Relativity

This is the discussion of "relativity" as known in Galileo Galilei's time.

Lorentz Ether Theory

This is the discussion of "ether" beliefs held by the Nobel prize winner, H. Lorentz.

Relativistic Mechanics

This is the study of mechanics using relativity equations and beliefs.

Actinometry

This is the technology of devices that measure radiation, radioactive elements, and so on.

Lava Dynamics

Lava is molten rock and is a kind of fluid. It is studied in fluid mechanics.

Kinetic Theory of Gases

This is a body of beliefs that gases are made of atoms in motion.

Flash Photography

This is a branch of photography about taking pictures of flashes and quick events.

Physics Museum Study

This is the study, design, and construction of museums dedicated to physics.

Casino Technology

Casinos are known as places where gambling occurs. Famous places are Las Vegas, Atlantic City, and Monte Carlo. This is the study of slot machines, roulette, and casino games and technology.

Fantasy Cosmology

Many ancient societies had "visions" of the universe like the World Tree, Cosmic Egg, Hindu turtles, Bindu, and Nun. This is the study and discussion of "cosmologies" that do not exist. It is derived from studying mythologies from around the world.

Physics "Pseudo-Equation" Study

Physics uses such famous equations as $F=ma$, $E=mc2$, $F=GMm/r2$, and so on. They are mathematical expressions with defined numbers called variables. However there are "expressions" that use math notation to form "pseudo-equations" (or false equations). An example is: Man + Woman + Love = Couple. It is a "toy-like" equation "invented" to seem like an equation and has no mathematics or physics value. Now, physics dramas can be "reduced" to expressions as pseudo-equations like "wind blowing". An example is "Air + Energy + Motion = Wind". Now try thinking about physics "processes" and phenomena and create your own pseudo-equations.

Acoustics

This is an immense science that studies sound. It has many sub-branches like audiology, infrasonics, supersonics, and ultrasonics. Its teachings are found everywhere from speakers to microphones to bells.

Stereophonics

This is a branch of radio technology studying speakers, stereo, and so on.

BRANCHES, SUB-DISCIPLINES, AND OFFSHOOTS OF PHYSICS

Ergodic Theory

This is a science that studies dynamical systems.

Multi-User Domain Study

Multi-user domains (MUDs) are "video gaming worlds" located on the internet. Many thousands of users access the MUD to live fantasies, go on adventures, and play around in "virtual worlds". They are a branch of video games known for intense gameplay, science fiction, fantasy, and captivating experiences.

SCUBA Technology

Scuba diving is the activity of diving with "aqualungs". Invented by such thinkers like Jacques Cousteau (of ocean exploration fame), it has grown to be an indispensable tool for diving worldwide. Its technology requires knowledge of physics and can be an intense study.

Artificial Life Study

This is the study of ways to create, tinker, simulate, and control living organisms.

Fluid Dynamics

This is a branch of physics studying the behavior of liquids (or fluids) like water.

Phenomenology

This is a branch of philosophy discussing phenomena and its behavior and meaning.

Question Study

There are six words called Who, What, Where, When, How, and Why. They form the basis of questions. Questions are vital to science. They are sentences asked to learn knowledge about things and to do science itself. Fundamental questions are questions asked of the nature of the universe. An activity in science is to compose as many questions

Interpretation Study

In quantum mechanics, there are issues called "interpretations" of quantum mechanics. Types are called the Copenhagen, many worlds, Penrose, and others. They are difficult issues to think about as the thinker is exploring the "meaning" of quantum mechanics. This is a subject taught in connection with quantum mechanics.

Granular Systems

This is the study of "grain-like" systems like sand castles, grain bins, and subjects involving atoms, graininess, and similar issues.

Telegraphy

This is a branch that studies telegraphs or a kind of telephone-like communication used in the 19th century. Topics here include Morse code, codes, and so on.

Tribology

This is the study of friction or collisions of surfaces and things.

Special Effects Study

In filmmaking, many moviemakers make science fiction movies that employ thrilling space scenes and other dramatic effects. The Star Wars movies are famous for their special effects. Topics here include stop motion, phaser battles, duels, space battles, and so on.

Condensed Matter Research

This is a branch of physics studying gels, frozen substances, crystals, novel states of matter, electronic devices, and other issues. It is described as the largest branch of physics. It is also called solid state physics as well.

BRANCHES, SUB-DISCIPLINES, AND OFFSHOOTS OF PHYSICS

Natural Philosophy

This is a subject that deals with asking questions about Nature. It originated before physics and physics in some sense had its beginning in this subject. Greek philosophy is a major topic here. Its discussions go back thousands of years and its impact on physics is immense.

Environmental Physics

This is a union of physics with environmentalism, ecology, and related subjects.

Archaeological Physics

This is a union of physics with archaeology. Topics here include mining, digging, ground penetrating radar, metal detectors, and so on.

Espionage Technology

This is the study of the tools of spies as used by the CIA and KGB.

Lithography

This is the study of making microchips, small computer components.

Egyptological Physics

This is a union of physics with Egyptology, the study of ancient Egypt.

Dwarf Planet Study

This is the study, search for, and analysis of dwarf planets or plutoids. There is a belief that a region of the solar system known as the Kuiper Belt hides many unknown dwarf planets. Its thought other unknown planets, perhaps the star Nemesis, and other objects lurk in this domain. In 2003, astronomer Robert Brown would discover the object named 2003 UB 313. It would ignite a famous controversy that perhaps a tenth planet was found. Later on this controversy would resolve where Pluto lost its planetary status and this object would gain the name of Eris.

THE GRAND SCIENCE OF PHYSICS

Wind Tunnel Technology
This is the study of wind tunnels.

Mountain Sculpting Technology
In South Dakota, there exist two mountains called Mt. Rushmore and Crazy Horse. One is carved with the likenesses of four US presidents by Gutzon Borglum and the other is being carved in the shape of the native American warrior, Crazy Horse. Carving mountains is a monumental activity and requires elaborate technology and discussion.

Physics Simulations, Computational Physics
This is the study of physical phenomena using computers.

Movie Physics
This is the study of films, special effects, and the film making process.

Mine Technology
This is the design and construction of mines, sea mines, minefields, and so on.

Physics T-Shirt Design
Many people like having T-shirts with Einstein, Maxwell's laws, and other physics arcana on them. It is a recreational hobby to design T-shirts with captions of physics on them.

Physics Web Design
This is the design of websites devoted to physics.

Science Outreach Centers
These are places (usually located in colleges) to instruct in physics or popularize science.

BRANCHES, SUB-DISCIPLINES, AND OFFSHOOTS OF PHYSICS

Physics Publications Study

This is the study, design, and activity of producing physics textbooks, magazines, and so on.

Violence Dynamics

This is the study of violent dramas like riots, lynchings, persecutions, acts of genocide, wars, etc.

Iconography

This is the study of statues, icons, and other constructions.

Da Vinci Study

Leonardo da Vinci was an epic artist, thinker, inventor, and dreamer of the Renaissance. He is known to have painted the Mona Lisa, wrote notebooks, and much more. His life and legend are something fascinating and are studied as a hobby.

Systems Theory

This is the study of complex systems.

Telemetry

This is the study of communication technology and related issues.

Paradigm Study

A paradigm is a "grand principle" in science. Its discussion tends to be intense in science.

Booby Trap Technology

Booby traps are "traps" to kill unsuspecting soldiers and are an issue in military science. Topics here include punjee traps, mine fields, aquatic mines, pits, and related issues.

Physics Poetry

Many people have taken to composing poetry, limericks, knock

knock jokes, anecdotes, and other "literary" creations discussing physics. These writings tend to be humorous, nonsensical, and witty and are a "past time" overall in physics. Issues to study here the Chinese philosophers, the Logicians, Zen Buddhism and Physics, Einstein quotations, and much more. Try composing your own verses in physics poetry.

Gravimetry
This is the study and measurement of gravitation fields on planets.

Geomagnetics
This is the study of the Earth's magnetic field.

Solar Power Satellites
This is the technology of having satellites in orbit. They receive solar radiation and convert it to energy to "beam" back to Earth for power.

Geochronology
This is the activity of studying the Earth's history and ages of life.

Levitation
Levitation is the notion of hovering or flying in the air or space without anything else. It is thought impossible or it could be because of antigravity. Issues like 'magnetic levitation' allow trains to be levitated via magnetic fields. So far levitation is a curious issue in physics.

Pharmaceutical Technology
This is the study of drugs, pharmaceuticals (drugs in pharmacy), and so on.

Sanitation Engineering
This is the discussion of how to make kitchens safe to work in and disinfecting places.

BRANCHES, SUB-DISCIPLINES, AND OFFSHOOTS OF PHYSICS

Spacesuit Design

This is the discussion of how to make space suits for astronauts.

Solar Storm Physics

This is the study of solar storms like prominences, coronal mass ejections, solar wind, etc.

Experimental Particle Physics, Accelerator Physics

This is the design and performance of experiments in particle physics.

LASIK Technology, Optical Medicine

This is a technology to cure eye problems affecting sight.

Zero Point Energy

It is thought that as matter is cooled to temperatures near Absolute Zero, then it is somehow possible to extract energy from space. So far attempts to do so have lead nowhere, but it remains as an enigmatic issue in energy production.

Nonlocality

Quantum mechanics is a vast and complex theory bewildering thinkers who explore it. There is an issue within known as 'nonlocality'. It is a discussion that perhaps particles have influences between them unknown to modern physics. It is a bewildering issue found in topics like Bell's theorem, quantum reality, and interpretations of quantum mechanics. So far it remains poorly investigated and bewildering to those who study it.

Agrophysics

This is the application of physics technology to agriculture.

Physics Biography Theory

Physics history knows epic thinkers like Einstein and Newton.

Their biographies are known throughout physics. In this subject, the physics student is asked to design his 'own biography' in the hopes he may be entered into physics history as a character of some kind.

Physics Chart Theory

A chart is like a map denoting things and qualities. In this subject, the physics student is asked to compose a chart showing physics knowledge ordered in a harmonious, compelling, and artistic way. He is to compose charts demonstrating each branch of physics.

Dirty Snowball Theory

This is the discussion that comets are "dirty snowballs" or asteroids of ice and rock.

In all physics is a vast and labyrinthine subject, more immense than anyone knows. Its teachings influence many other domains and an education in it is required to understand many other subjects. Physics as of now is continually expanding and diversifying of its results, subjects, developments, technologies, and advancements. It has now grown so complicated it seems as if no can keep up with it anymore. Physics is a labyrinth of a science, daunting and complicated, a morass of subjects more than just what physics is.

"Knowledge is infinite, you cannot know it all. Leave it to God to know everything."

-Voigt

In the next section, we will now review a discussion on the Nobel Prize for Physics.

Chapter 6
The Nobel Prize

"Society glorifies the explorer, the athlete, the famous politician, rock stars, and movie stars. For some reason it ignores the scientist."

-physics lecture

The Nobel Prize for Physics

In this section, we will explore a discussion on an award that has become known as physics' most celebrated honor. It was begun by the dynamite inventor Alfred Nobel and has since acquired a prestige beyond Nobel's intentions. Many physicists celebrate that they have won the Nobel prize and Nobelists are accorded honor akin to royalty for their epic achievements. The Nobel prize is today one of the world's greatest science honors honoring men and women who did great acts for physics and science overall. Presented here is a discussion on the Nobel prize for physics.

Alfred Nobel

In the 19th century, there lived an inventor named Alfred Nobel from Sweden working in explosives. He developed a new kind of explosive called "dynamite" which was an improvement over existing explosives. He marketed his product and became very rich over it. However, dynamite was used by people for all sorts of purposes. Examples include mining, burrowing, digging canals, and "positive" uses. However, others used his explosive for robbing banks, murder, warfare, acts of terrorism and wanton violence, assassinations, developing new weapons, making land mines, and other "negative" uses. He was "outraged" by this "misuse" of his invention that it disturbed his conscience so. He wrote a will outlining the creation of a series of "prizes" to commemorate "acts of goodness on behalf of the welfare of Man". In his will, he wrote his prizes should be given to people who have done outstanding work in the fields of physics, chemistry, medicine or physiology, literature, and world peace (later a prize for economics was added). His prizes were to be financed from his personal wealth and were to be awarded after his death. After he died, there was a "struggle" to cancel the prizes and give his wealth to his heirs, however the prizes were founded and were first awarded in 1901. Today, the prizes named for Alfred Nobel (the Nobel Prize) is an international symbol of grand achievement in science. They have become the "highest" honor in science with a mystique of genius all their own. Since then, the tradition of the Nobel prizes continues honoring new people for their great acts to science and other fields as directed by Nobel. Nobel prizes are awarded yearly and to no more than three recipients who are alive. Recipients are honored without bias to nationality, age, religion, or race. This is done to maintain the "mystique", rarity, and "special occasion" on being chosen a Nobel laureate. Today, the Nobel Prize for Physics is considered physics' greatest award for physics research, discovery, and achievement.

The reader is encouraged to explore more indepth books on the Nobel prize and the life of Alfred Nobel.

Nobel's Will

Nobel in his lifetime made a fortune from the sale of dynamite. He amassed millions of dollars that made him one of the world's wealthiest men in the 19th century. Nobel wrote a will (a document stating how his fortune was to be spent after his death) about what to do with his money. He stated that his money be used for the creation of prizes for scientists, writers, and peacemakers. His family tried to challenge his will so they could have his money. Their efforts failed and thus the Nobel Prize was created.

Nobel Institute (Nobel Museum, Nobel Foundation)

This is a famous institute that oversees the activity of researching and awarding the Nobel Prize. It maintains offices, a museum, and relations with academies. It oversees committees who research and select yearly Nobel prize winners and administers Alfred Nobel's will.

Nobel Committee

Each year, committees of eminent persons convene in Sweden and Norway. They receive submissions from past Nobel prize winners, physics professors, and other people. They are given the names and achievements of many scientists and then they proceed to investigate, sort, and judge from the various submissions as to what is "prize winning work". Committees work in "secret" and are not in the service of any government. They are tasked with selecting no more than three people who are alive and who participated in some science drama. They then chose who the "Nobel prize winners" of the following year will be. They work in connection with three bodies called the Royal Swedish Academy of Sciences, the Nobel Foundation, and Karolinska Institute.

Nomination for the Nobel Prize

The Nobel prize is physics' and science's "highest award" known worldwide for eminence and prestige. To be nominated for a Nobel prize requires that scientists have done something outstanding for science. This means they participated in a discovery, developed a technology, lead a science team, or did something monumental in physics. Often times, scientists may consist of teams who discover something. However the Nobel prize will only commend up to three people and no more. Thus if more than four people participate in a Nobel prized drama, then extra persons tend to be bypassed and this creates "scandal" over awarding the Nobel prize. For the most part, science work must be tested by the years, become respected by professional scientists, and win the support of Nobel prize nominators in being chosen as possible Nobel caliber work. It is thought to be a great honor to be nominated for a Nobel prize as this shows that a scientist has done great work. However, the Nobel prize can only be awarded once a year and thus a lot of Nobel caliber work will unfortunately never be commended with a Nobel prize.

The Nobel Prize Ceremony

The Nobel prize is awarded in grand royal galas in Stockholm, Sweden every December of every year. The "prize" is awarded personally by the King of Sweden in staged ceremonies in royal attire. The prize ceremony is an annual drama attended by many dignitaries of nations. It is the premier "social event" of Sweden known for speeches, royal balls, lavish excess, and the solemn occasion of commemorating the life of Alfred Nobel, his prizes, and the great acts of science honored here. The "laureates" (prize winners) give speeches explaining their life, careers, and achievement. They are then entered into official records and are given a "cash gift" of about a million dollars. Nobel laureates are given money, citations, and a gold medal with Alfred Nobel's image on it for their grand achievement. "Nobel week" as it is called usually lasts only a week and laureates gather from around the world to attend the ceremony of the Nobel prizes. Later on, Nobel laureates are "enshrined" in a museum commemorating the Nobel prize and their great acts of science. In the prize ceremony itself, laureates will sit on a stage awaiting to meet the King of Sweden. Here the king will give the Nobel prize to the laureate after which there will be applause. In itself it is a grand occassion celebrating science, giving awards, and staging an epic drama for guests, royalty, and of course winners of the Nobel prize.

Nobel Lecture

Many winning Nobel laureates go on to give lectures describing their biography, the work for which they won a Nobel prize, and so on. They may appear at lectures during "Nobel week" and "explain" their life and work. For many people, Nobel prize-winning work in science is a confusing thing and needs to be explained. Various Nobel laureates have written books trying to explain their work even though much of it is garbled by scientific jargon and advanced concepts.

Nobel Citation

A Nobel citation consists of a gold medal with Alfred Nobel's image on it, a diploma, a cash grant, and a statement for which the Nobel prize is awarded. The Nobel citation is given to every Nobel laureate and is a kind of "souvenir" at having won the Nobel prize. Nobel citations tend to be displayed in a museum or prominent place. People who have won the Nobel prize are entered into records like almanacs and history books.

The Nobel "Phonecall"

Every October of every year, the various Nobel committees conclude their discussions and select the Nobel laureates of that year. They will stage press conferences announcing who has won and for what for (the citation). They will "phone" the prize winning scientists from around the world that they have won the Nobel prize. To the people who win a Nobel prize, this drama comes as a "shock" whereby they are told of their commendation. After that, a press conference may be arranged and the press shows up to interview them. Celebrations may commence whereby colleges honor them, champagne is drank over the occasion, and a party atmosphere may erupt. Suddenly, the winning scientists become "celebrities" as their Nobel citation is publicized worldwide. In all, the Nobel October week is a much anticipated drama where many scientists eagerly await the news if they have won the Nobel prize or not.

Nobelitis

Many scientists around the world have done Nobel prize-quality work. Many of them are nominated for the prize, however they may only know by rumor that they have been nominated. Many scientists wait expectantly each year wondering if they have won the Nobel prize. This "anticipation" has created a condition whimsically referred to as "Nobelitis" (or the disease of anticipation as to whether one has won the Nobel prize or not). Many physicists are known to "suffer" from this "anxiety".

Nobel Tradition

The Nobel Prize has been given since 1901. It began with Nobel's will and continues on even now. At times, it was suspended due to world war anarchy, but was later revived. Nobel specified that there be only five prizes and no more. However, there was an effort to make a Nobel prize for economics and this was successful. As of now, there is a movement to reject the creation of more Nobel prize categories. Today, Nobel laureates are given cash awards of well over a million dollars. If a Nobel award commends three people, these three split the award between themselves.

Nobel Prize Winning Work

The Nobel prize for physics is given for work, discoveries, and achievements that celebrate a dramatic and "important" contribution to physics. It is an award that celebrates the science and the scientists who found something great in physics. Nobel prize winning work generally consists of these qualities:

- Epic experiments that shaped physics, changed notions, tested theories, found something great, or discovered something fundamental about the universe.
- A technology that shaped physics or science in some great way.
- The discovery of something thought to be "great" and "fundamental" to physics like:
- New energies –States of matter –New particles –New forces –New effects
- The act of unifying forces of Nature.
- The proposal of a theory or some other work judged "great" in physics history.
- A lifetime of epic service to physics with many important discoveries.
- The act of overthrowing a cherished belief in physics.
- The discovery of a new law of physics.
- Pioneering new branches of physics, areas of study, or other "frontier" issues.
- Advancing a domain of physics in "incredible" ways.
- Advancing understanding of particle physics.
- Advancing understanding of superconductivity.

THE NOBEL PRIZE FOR PHYSICS

The Nobel Prizes Given

The Nobel prize for physics was first awarded in 1901 and has since been a "tradition" of science ever since. Each year, new "laureates" are invited to Sweden to receive the prize and it is today a much awaited yearly event in the science. In this section, we will review what past Nobel prizes for physics have been, who received them, and what were they given for.

1901 In this year, Wilhelm Roentgen was honored for discovering X rays. X rays are a radiation (or energy) that can "see" through mass bodies and are used heavily in medicine. X rays have since gone on to be a legendary discovery inspiring many kinds of technology. X rays have been used in medicine, astronomy, and in many other domains. As a sidenote, the fictional superhero, Superman is credited with X ray vision or the power to see through clothing and barriers "magically".

1902 Hendrik Antoon Lorentz and Peter Zeeman were honored for their work in the Zeeman effect. This effect discusses how electric fields can influence a spectrum (or a collection of colors like red, orange, and so on). Lorentz would live an epic life in physics influencing relativity and other domains.

1903 Antoine Henri Becquel, Pierre and Marie Curie were honored for radioactivity, the decay of atoms in radiation displays. Radioactivity comes in varieties like alpha, beta, and gamma and it can cause cancer if "improperly" handled. Radioactivity is the behavior of atoms to release energy and particles as their nuclei self-destruct or decay. Marie Curie is famous for being a "grand lady" of physics for winning two Nobel prizes and shaping modern physics. Marie Curie is honored with an element, curium, the Curie Institute, and so on.

1904 John William Strutt, Baron Rayleigh was honored for his work in Argon (a gas), Rayleigh scattering, etc.

1905 Philip Lenard is honored for his work on the cathode ray tube (or CRT). The CRT would go on to lead to television, computers, and video games shaping the modern world.

THE GRAND SCIENCE OF PHYSICS

1906 Sir Joseph John Thomson (JJ Thomson) discovered the electron. The electron is a particle with a negative charge and is a vital part of particle physics. It is central to the understanding of atoms, chemicals, subatomic particles, and much more. The movement of electrons is called electricity. The electron is considered the source of this energy that has revolutionized technology in appliances.

1907 Albert Abraham Michelson and his work in measuring the speed of light. The speed of light is considered a fundamental and important number to know and is vital to physics. Michelson participated in the legendary Michelson-Morley experiment which showed "ether" did not exist. This result lead to relativity by Einstein later on.

1908 Gabriel Lippman was honored for his work in color photography. Lippman's work changed photography by making black and white images have color and is used seemingly everywhere. His work is felt in color photographs found in magazines, books, and movies.

1909 Guglielmo Marconi and Ferdinand Braun. This duo worked on pioneering "radio" (called "wireless telegraphy"). Radio went on to become radio stations, radio astronomy, radar, and so much more, a great invention.

1910 Johannes van der Waals. He is known for the discovery of "van der Waals" forces important in physics and chemistry.

1911 Wilhelm Wien. He discovered the famous Wien's law (veenz law), a finding in heat. This finding would lead to quantum theory.

1912 Nils Dalen. He did important work in lighting technology. His work lead to acetylene torches used in welding.

1913 Heike Kamerlingh Onnes. He discovered superconductivity. This issue is about "cooling" masses to temperatures near Absolute Zero, the coldest possible temperature. Masses allow electricity to flow without resistance, this is thought to have revolutionary importance in industry. Today, research into superconductivity is immense in physics with more Nobel prizes given over advances in its issue.

1914 Max von Laue. He is known for his work in X-ray diffraction, crystals, and X ray spectrums.

1915 Sir William Bragg (both Senior and Junior, father and son). They did important work in X ray crystallography. X ray crystallography is the act of passing X rays through crystals to obtain photographs. From there, the structure of chemicals can be found.

1916 No award given, possibly due to World War 1 chaos.

1917 Charles Barkla. He did important work in X ray scattering. His work was an important offshoot of the discovery of X rays earlier on.

1918 Max Planck. He discovered the "quantum" or the belief that energy comes in "packets". It lead to the quantum theory (quantum mechanics) and is hailed as a revolution in physics. Planck went on to live a life as a physics "superstar" and he is commemorated in organizations, medals, the concept of Planck time, the Planck Institute, and the Planck's constant.

1919 Johannes Stark. He is known for the Stark effect which would lead to quantum theory.

1920 Charles Guillaume. He is known for his work in alloys of nickel (a kind of metal). He invented metals called invar and elinvar useful in technology.

1921 Albert Einstein. He is honored for his work in the photoelectric effect. This effect discusses how light impacts metal and produces electricity. This finding inspired solar cell technology used worldwide. Einstein lived a monumental life in physics and has many biographies. It is thought overall that Einstein deserved many Nobel prizes and research he worked on continues to be main research in physics today. Today, Einstein is physics' most famous personality and his life and legend are found throughout the world. At the time of the prize, it was believed relativity was just too controversial of topic and thus no prize was ever given for relativity. Einstein would later give his Nobel prize money to his ex-wife, Mileva Maric. There is a controversy that Maric worked with Einstein in his discoveries and thus deserved the Nobel prize too, she would not be awarded the Nobel prize.

1922 Niels Bohr. He is honored for his work in the Bohr model, a model of atoms. Bohr went on to be a living legend in physics in

Einstein's time. He founded the Niels Bohr Institute and had a stature similar to Einstein in physics.

1923 Robert Millikan is famous for the "oil drop experiment" measuring the electronic charge constant. Electronic charge is the amount of charge on an electron. The oil drop experiment would become one of physics history's classic experiments.

1924 Karl Siegbahn. He did important work in X ray spectroscopy. Spectroscopy is about studying the "colors" of a kind of radiation and then determining the identity of chemicals.

1925 James Franck, Gustav Hertz. This "team" did important work in electrons and the Bohr model. The unit of frequency is named the "hertz" in Hertz's honor.

1926 Jean Perrin. He is known for his work in the discontinuous nature of matter. His work would influence many physics branches and Perrin lived a productive life doing grand things in science.

1927 Arthur Compton, Charles Wilson. This duo are famous for the Compton effect and advances in detectors of particles.

1928 Owen Richardson. He explored thermionics, a branch of physics dealing with heat technology. This would lead to the vacuum tube and ultimately computers.

1929 Louis de Broglie. He discovered "matter waves" which shows matter has both wave and particle qualities. This was later used in electron microscopes, a kind of microscope known for seeing very small objects.

1930 Sir Chandrasekhara Raman. He is known for Raman scattering and insights into light's nature.

1931 No award given.

1932 Werner Heisenberg. He is known for the Uncertainty Principle. The Uncertainty Principle is a belief in physics about how difficult it is to measure quantities in quantum mechanics. It is highly technical and requires a first year course in quantum physics to more thoroughly explain. Heisenberg is known for being one of the great thinkers of quantum mechanics. He is infamously known for working for Adolf Hitler and running the Nazi atomic bomb project.

THE NOBEL PRIZE FOR PHYSICS

1933 Erwin Schrodinger, Paul Dirac. This duo is famous for their pioneering work in quantum mechanics. Schrodinger is famous for his contributions like Schrodinger's wave equation (described as a very difficult subject to know) and the Schrodinger's Cat (not a real cat, but a thought experiment discussing a cat that can both be alive and dead). Dirac is famous for his Dirac equation, a finding in quantum mechanics, the "Dirac sea", and predicting antimatter. Both these men lived legendary lives in physics.

1934 No award given.

1935 James Chadwick discovered the neutron. The neutron is a central particle of the nucleus. It has no charge, but has the mass like a proton. The neutron is vital in atomic bombs, particle physics, and related subjects.

1936 Victor Hess and Carl Anderson discovered cosmic rays and antimatter.

1937 Clinton Davisson and George Thomson. They investigated diffraction in crystals. Their work would be found in the Davisson-Germer experiment and other dramas.

1938 Enrico Fermi. He discovered new elements and helped discover fission or the effect allowing an atomic explosion. Fermi lived an epic life in physics with many biographies. Famous episodes of his life are the Manhattan Project, the first nuclear reactor at Stagg Field, the Fermi Paradox, fermions, and so much else. The element 'fermium' is named for him.

1939 Ernest Lawrence. He pioneered the cyclotron. The cyclotron "pioneered" the use of particle accelerators of "atom smashers". Many grand installations derived from his work were made. Famous ones are SLAC, CERN, Fermilab, and many others. He essentially began the subject of particle physics. Lawrence is honored in the laboratory called Lawrence-Livermore and the element, lawrencium.

World War 2 would strike Europe in 1940. Nazis would invade Norway and much of Europe. While fighting was intense, the Nobel prizes were suspended for this time until peace could return later on. Adolf Hitler would forbid Germans from receiving the Nobel prize.

THE GRAND SCIENCE OF PHYSICS

1940 No award given. Possibly due to World War 2 anarchy at the time.
1941 No award given. Possibly due to World War 2 anarchy at the time.
1942 No award given. Possibly due to World War 2 anarchy at the time.
1943 Otto Stern. He worked on the molecular ray method. He contributed to the famed Stern-Gerlach device.
1944 I. I. Rabi. He investigated the atomic nucleus.
1945 Wolfgang Pauli discovered the Exclusion Principle, a belief that particles cannot occupy the "same quantum state". Pauli is known as a "dandy" in physics for his epic achievements and for living a "great life" in physics. In his time, he was a celebrity like Einstein participating in many grand dramas of physics.
1946 Percy Bridgman worked on pressure chambers.
1947 Sir Edward Appleton investigated the ionosphere (the Appleton layer), a layer of atmosphere known for particles, energy, and electricity.
1948 Patrick Blackett worked on cloud chambers, a machine that detects particles. Blackett also sought for a grand unified theory and is known to have failed.
1949 Hideki Yukawa worked on mesons (the strong nuclear force), a kind of particle that affects the nucleus and the force inside.
1950 Cecil Powell worked on photographic methods to detect subatomic particles.
1951 Sir John Cockroft and Ernest Walton worked on particle accelerators. A classic invention of theirs is the Cockroft-Walton particle accelerator.
1952 Felix Bloch and Edward Purcell worked on "magnetic resonance imaging". This would be used in medicine to detect cancer and is found in many hospitals.
1953 Fritz Zernike worked on the phase contrast microscope, a fascinating kind of microscope for seeing incredibly small objects.
1954 Max Born and Walter Bothe did important work in quantum mechanics. Born was a giant of quantum mechanics and wrote many textbooks.

THE NOBEL PRIZE FOR PHYSICS

1955 Willis Lamb and Polykarp Kusch did work on the Lamb Shift and magnetic moment of the electron. These findings would have great influence in physics.

1956 William Shockley, John Bardeen, and Walter Brattain invented the transistor. The transistor would go on to be a "great invention" helping begin the computer revolution, leading to the transistor radio, and inspiring other kinds of technology.

1957 CN Yang and TD Lee for their work on parity. Parity is the idea that atoms "decay" in their nuclei according to a specific direction. When "parity violation" was discovered, it caused a shock that swept physics as it overthrew a long cherished belief in physics. The Nobel committee hastily awarded them a Nobel prize for this epic finding. Both Lee and Yang are from China and they are celebrated as great scientists in this country.

1958 Pavel Cerenkov, Ilya Frank, and Igor Tamm for their work on Cerenkov radiation. Cerenkov radiation is a "blue light glow" found in atomic reactors and is a technical issue.

1959 Emilio Segre and Owen Chamberlain discovered the antiproton. The antiproton is the antimatter counterpart of the proton. Many physicists have used this particle to make antimatter elements, a curiosity in physics.

1960 Donald Glaser invented the bubble chamber, a kind of particle detector.

1961 Robert Hofstadter and Rudolf Mossbauer worked in resonance and the Mossbauer effect.

1962 Lev Landau worked in superfluids. A superfluid is a super-cold liquid that flows up hill and without resistance. Its investigation is considered a major area of physics research. Landau himself was a giant of physics in the Soviet Union.

1963 Eugene Wigner, Maria Mayer, and J Hans Jensen worked on atomic models or pictures of how the atom behaves or looks like.

1964 Charles Townes, Nicolay Basov, and Alexander Prokhorov worked on lasers and masers. These inventions would lead to laser welding and laser technology of all kinds.

1965 Sin-Itiro Tomonaga, Julian Schwinger, and Richard Feynman worked on quantum electrodynamics. This theory was an improvement on how to understand how electricity behaves. Feynman lived a life as a "physics adventurer" for his work, his personality, the famous Feynman Lectures on Physics, his writings, Feynman diagrams, investigating a space shuttle crash, his work on the Manhattan Project, and more. Biographies on him tend to be best-sellers and his writings are widely known. Both Schwinger and Tomonaga went on to live epic lives in physics.

1966 Alfred Kastler worked on resonance methods. His work would greatly influence lasers.

1967 Hans Bethe worked on fusion in stars. He explained the age old mystery "Why does the Sun and the stars shine?" Bethe (bay tay) lived an epic life in physics and has many biographies over his adventures.

1968 Luis Alvarez worked on the bubble chamber, a kind of particle detector. Alvarez is famous for proposing an asteroid hit the Earth caused the extinction of the dinosaurs.

1969 Murray Gell-Mann worked on quarks or particles that make up protons and neutrons. Curiously, others were working on issues like Gell-Mann's and its thought others (like George Zweig) deserved a Nobel prize for quarks as well.

1970 Hannes Alfven and Louis Neel worked on magnetohydrodynamics. This subject deals with how magnetic fields influence plasmas (ionized gases). Alfven would become famous for Alfven waves and ambiplasma cosmology.

1971 Dennis Gabor worked on holography. Holography is about making photographs that have images in three dimensions (trick photographs). Holography is used in many new kinds of technology and is evolving still.

1972 John Bardeen, Leon Cooper, J. Robert Schrieffer worked on the BCS theory, a theory of superconductivity. BCS theory in reality stands for Bardeen-Cooper-Schrieffer theory on superconductors. It was a theory that offered a new explanation for how these chemicals behave with electric currents passing through them. Bardeen is a legend in physics for winning two Nobel prizes.

1973 Leo Esaki, Ivar Giaever, and Brian Josephson worked in electronics technology.

1974 Sir Martin Ryle and Anthony Hewish discovered pulsars. Pulsars are neutron stars. At first, they were thought to be radio signals from aliens, but were later identified as an astronomical object. A graduate student named Jocelyn Bell shared in the pulsar's discovery and is famously known to be overlooked in receiving a Nobel prize over this.

1975 Aage Bohr, Ben Mottelson, and Leo Rainwater worked on atomic structures. Aage Bohr is a son of the famed physicist, Niels Bohr.

1976 Burton Richter and Samuel Ting discovered the J/psi particle. This caused an event in physics dubbed the November Revolution. Both Richter and Ting lead scientist teams that independently found a new kind of subatomic particle that showed "charm" existed. One team called it the psi and another called it the J. As a compromise, this new particle was jointly named the J-psi in honor of them both.

1977 Philipp Anderson, Sir Nevill Mott, and John Van Vleck worked in "magnetically disordered systems". Their work is highly technical and inspired many kinds of technology.

1978 Pyotr Kapitsa, Arno Penzias, and Robert Wilson. Kapitsa worked in superfluids and low temperature physics. Penzias and Wilson found the cosmic background radiation. Please study the story of the cosmic background radiation as it was a monumental discovery about the nature of the universe. It settled a long discussed debate about the origin of the universe.

1979 Sheldon Glashow, Steven Weinberg, and Abdus Salam found the electroweak theory. Electroweak theory is the unification of the weak nuclear force and electromagnetism, both fundamental forces of physics. Both Glashow and Weinberg would live epic lives in physics while Salam would found an institute.

1980 James Cronin and Val Fitch for their work on the CPT theorem. This issue is similar to parity violation and its discovery stunned physicists worldwide. This was a followup to the famous Lee and Yang parity drama.

1981 Nicolas Bloembergen, Arthur Schawlow, and Kai Siegbahn for their work in high resolution spectroscopy. Schawlow contributed to laser technology.

1982 Kenneth Wilson worked on critical phenomena in phase transitions.

1983 Subrahmanyan Chandrasekhar and William Fowler for their work in astronomy. Chandrasekhar is famous for a "limit" named for him. Both men lived famous lives in astrophysics for shaping the subject along with Fred Hoyle (of steady state cosmology fame).

1984 Carlo Rubbia and Simon van der Meer for discovering the W and Z particle. The W and Z particles are key predictions of electroweak theory that won a Nobel prize in 1979.

1985 Klaus von Klitzing for his work in the quantized Hall effect.

1986 Ernst Ruska, Gerd Binning, and Heinrich Rohrer worked on electron microscopes and the scanning tunneling microscope. Ordinary microscopes use light to see objects like cells. These men built microscopes that could see small objects using electrons and the tunneling effect, an obscure issue in quantum mechanics.

1987 J. Georg Bednorz and Karl Muller worked on superconductivity. They caused a sensation finding superconductivity in ceramics (clay like substances) causing an event called the "Woodstock of Physics". It is known superconductors (or substances that conduct electricity without resistance) only exist at temperatures near Absolute Zero. It is a "dream" of this subject to make superconductors near temperatures like 20 degrees Celsius (or room temperature). Efforts are underway to find chemicals that will do just that but so far without success. However sensations do occur when substances are found to be superconducting at temperatures well above Absolute Zero.

1988 Leon Lederman, Melvin Schwartz, and Jack Steinberger for their work on neutrinos. Neutrinos are very small particles with no electric charge.

1989 Norman Ramsey, Hans Dehmelt, and Wolfgang Paul worked on atomic clocks and ion traps. Atomic clocks are known as some of the world's best clocks for using atomic decay to power a clock.

1990 Jerome Friedman, Henry Kendall, and Richard Taylor worked on quarks.

1991 Pierre Gilles de Gennes worked on liquid crystal polymers. This finding would lead to novel kinds of screens used on TV sets, calculators, and other technology.

1992 George Charpak worked on the multiwire proportional chamber, a kind of particle detector.

1993 Russell Hulse and Joseph Taylor studied a binary pulsar (or two pulsars orbiting each other) and discovered gravity waves. Gravity waves were a prediction by Einstein for decades, but were discovered by this pair. They opened the field of gravitational astronomy. Today gravity waves are a good topic in astronomy. Detecting them is extremely difficult. It is thought the Big Bang generated gravity waves as well. Detecting them may lead to insights into the Big Bang.

1994 Bertram Brockhouse and Clifford Shull worked in the technology of neutrons.

1995 Martin Perl and Frederick Reines discovered the tau lepton, a kind of subatomic particle.

1996 David Leel, Douglas Osheroff, and Robert Richardson worked in superfluids.

1997 Steven Chu, Claude Cohen Tannoudji, and William Philips worked on laser cooling. Laser cooling is the technology of using lasers to cool matter to temperatures near Absolute Zero. Chu would go on to serve as Secretary of Energy under US president Barack Obama.

1998 Robert Laughlin, Horst Stormer, and Daniel Tsui worked on the fractional quantum Hall effect.

1999 Gerard t'Hooft and Martinus Veltman worked on renormalization. This is a mathematical trick that allows for handling calculations in quantum mechanics.

2000 Zhores Alferov and Herbert Kroemer worked on optoelectrics (or electronic devices using light). Jack St. Clair Kilby worked on the integrated circuit. The integrated circuit would influence computers.

2001 Eric Cornell, Wolfgang Ketterle, and Carl Wieman created the Bose-Einstein condensate. The BEC is a new kind of matter predicted by Einstein and Bose and their work lay neglected for over 70 years before Cornell, Ketterle, and Wieman came along trying to make their new kind of matter. The BEC is whimsically also called the "superatom" and is a highly technical issue.

2002 Raymond Davis, Jr., Masatoshi Koshiba, and Riccardo Giacconi worked on the solar neutrino problem. This was a famous long standing problem in astronomy. It essentially says that the Sun was not emitting neutrinos in the quantities that it should.

2003 Alexei Abrikosov, Vitaly Ginsburg, and Anthony Leggett worked in superfluids and superconductors.

2004 David Gross, David Politzer, and Frank Wilczek worked on asymptotic freedom. Literally meaning "freedom from the line", it was a finding crucial to understanding quarks.

2005 Roy Glauber, John Hall, and Theodor Hansch worked in optics technology.

2006 John Mather and George Smoot worked in cosmology using the COBE satellite. They studied an issue called the "Dark Ages" and found evidence for the Big Bang. They explored an era of the universe previously only predicted. Today, Smoot and Mather are stars in astrophysics for leading science teams that worked with the COBE satellite.

2007 Albert Fert and Peter Grunberg worked in giant magnetoresistance. Called GMR, this issue lead to all kinds of technology found in hard drives in computers.

THE NOBEL PRIZE FOR PHYSICS

2008 Yoichiro Nambu, Makoto Kobayashi, and Toshihide Maskawa worked in broken symmetry, quarks, and particle physics. All these men hail from Japan and Nambu is famous for his work in particle physics. There is an on going controversy about physicist Nicolo Cabibbo of Italy having done work related to them, but not getting Nobel recognition. Nambu would be most famous for naming the Nambu-Goldstone boson, a kind of unknown particle.

2009 Charles K. Kao, Willard S. Boyle, George E. Smith. Kao is known for beginning the issue of fiber optics or passing light beams through glass wires. Boyle and Smith invented the charge coupled device, a machine famous in photography and detectors. These two inventions are both famous for being among some of the greatest in physics history.

2010 Andre Geim, K Novoselov. These men worked in graphene, a technical subject.

2011 Saul Perlmutter, Brian Schmidt, Adam Riess. Cosmologists would discover by studying supernova that the universe is undergoing a cosmic expansion. It was discovered that the cosmic expansion was somehow speeding up or 'accelerating' and this is currently unexplained.

2012 Sergei Haroche, David Wineland. A quantum computer is a kind of computer able to process information much faster than any 'classical' computer. So far it has not been developed. However Haroche and Wineland would make discoveries that would advance the field of quantum computers.

Please read about the various Nobel prizes of the past and the epic moments in physics they commemorate.

As of now the Nobel Prize is a grand tradition of science and continues to being awarded. It is unknown how long it lasts as a tradition, but its drama defines modern science and overall the modern world as well in science, technology, and the recognition of scientists.

Other Nobel Prizes

Its known many physicists have won Nobel prizes in other fields for doing important work. Some physicists have gone on to win the Nobel prize for chemistry and the Nobel peace prize. Famous names here are Ernest Rutherford (discovering the nucleus), Otto Hahn (fission), and Andrei Sakharov (builder of the Soviet hydrogen bomb). Many peoples and organizations would know that the Nobel Prize only commends six categories of sciences and topics. For this other groups would found kinds of prize to commend thinkers in other domains like mathematics, geology, psychology, and so on. It would lead to the creation of prizes that would try to mimic the Nobel Prize for its glory, mystique, and to correct for its faults of not recognizing scientists in other domains. Other prizes that try to do this are the Wolf Prize, Vetlesen Prize, Crafoord Prize, and Abel Prize.

Consequences of the Nobel Prize

Winning the Nobel prize has been described as a "great moment and high honor" for those lucky enough to be given it. It is a statement of achievement that certain scientists have done something momentous and ground-breaking for the "good" of physics. It is a drama that celebrates the physicists who found or did something great in physics, their achievement, and physics in general. Winning the Nobel prize is described as a "dream" or "fantasy" in physics. It accords a "celebrity" to physicists who win one that they are accomplished scientists and luminaries of the subject. The Nobel prize is identified with Einstein, Bohr, De Broglie, Lawrence, Rutherford, Thomson, Feynman, and many others who became legends of physics and with acts of genius overall. It is a rare honor and many nations and colleges celebrate Nobel prize winners among their number. The Nobel prize takes otherwise "unknown" professors and changes them into "stars" of science and people of reknown. The Nobel prize can change people's lives with new opportunities, speaking engagements, and all kinds of dividends. Nobel prize winners are considered "eminent colleagues" in physics on "equal" terms with their undistinguished colleagues. It is an international honor that has a mystique and fame beyond the profession of physics.

Nobel Prizes of the Future

The Nobel prize is given for outstanding and monumental acts and discoveries of physics. Each prize known to history celebrates a monumental achievement in some way. In this section, we will explore issues that perhaps could win a Nobel prize for physics in the future. While it is uncertain if Nobel prizes will actually be given for these achievements, it is quite possible they will gain consideration as prize worthy. There are in physics many mysteries and curious issues awaiting physicists to come investigate and solve these problems. These issues have a "momentousness" or "significance" that they are in some way a "major" problem in physics. The Nobel prize commemorates "great" acts of physics and it is subjective that these could be described as "great", however these mysteries have a fascination that they could be the harbinger of future physics chapters to be. Discussed here are issues on the frontier of physics. No one knows where they will go or if they will be solved of their mysteries. But they beckon to future ages of explorers into the frontier of physics. Examples of issues that may (only speculation) win Nobel prizes in the future are:

- Unruh radiation (a speculated kind of radiation by physicist, William Unruh.) This is a speculated radiation thought to exist for observers if they undergo acceleration. It is not known to exist and remains as an issue in physics.
- Higgs boson (a particle postulated by Peter Higgs for giving mass to particles.) In advanced physics, there is a belief in a particle that could give mass to particles. It is named for the thinker Peter Higgs and is a prediction of the famed Standard Model of physics theories. News reports in 2012 state it may have been found at the particle accelerator facility, CERN. So far it remains and enigmatic issue and possibly a Nobel Prize will be given over it.

THE GRAND SCIENCE OF PHYSICS

- The discovery of the "string", supersymmetric particles, the chargino, Higgsino, neutralino, axino, Wino, Zino, goldstino, and other "mythical" subatomic entities. Particle physics is filled with many kinds of predictions of particles. While many have been found, many more remain unresolved.
- Goldstone boson (a kind of particle found in advanced physics discussion.)
- The Oh My God Particle is an unusual kind of particle from space known for extremely high energies. It is poorly investigated and could lead to astronomical work.
- Hawking radiation (Stephen Hawking theorized that black holes may decay.)
- Black holes (these are astronomical objects known for intense gravity and discussion.)
- Mini-black holes. This refers to black holes the size of atoms or nuclei.
- Proton decay (there is discussion that the proton may decay into smaller particles.)
- Neutrino mass problem (its thought the neutrino may have mass deciding the fate of the universe.)
- Rotating universe (this is the belief the universe is some how spinning as it expands.)
- Solution to the mysteries of cosmology (cosmology is filled with many mysteries as of now like dark matter, inflation, the Big Bang, the horizon problem, the end of the universe, and so on.)
- Issues like inflation, eternal recurrence, the cyclic model, the horizon problem, the hierarchy problem, and the flatness problem.
- Breakdowns of quantum mechanics or relativity.
- Issues like nonlocality, Bell's theorem, quantum reality, quantum teleportation, etc.
- Theory of quantum gravity (there is an effort to join relativity with quantum mechanics.)

- Grand unified theory (physicists want to find this ultimate theory of Einstein dream.)
- New theories of superconductivity (Its thought the BCS theory will be replaced someday.)
- Multiverse. Its believed the universe is alone and unique as an entity. However there are speculations that other universes exist and that the universe itself lays within a grander, infinite reality of other universes (called the multiverse).
- Planck dimensions (At temperatures and pressures near the Big Bang, its thought fundamental new physics will be revealed.)
- New dimensions of space (It's a current fad that space may have more than four dimensions.)
- The creation of new branches of physics.
- The combining of physics with another science or issue.
- Magnetic monopoles (Its thought lone magnetic pole particles may exist.)
- The discovery of astronomical objects called the Great Attractor and Great Wall.
- Discoveries from the Large Hadron Collider at CERN.
- Tachyons (Its thought faster than light particles may exist, but cannot be found.)
- Preons, smallestons (this is a belief that particles smaller than quarks exist.)
- Anti-galaxies (this is the belief that galaxies of antimatter may exist.)
- The study of a new kind of phenomena called the "bosenova".
- The discovery of the "black ring", a speculated phenomena showing higher dimensions exist in a particle accelerator.
- –Baby universes (this refers to ideas that new universes may arise and vanish in spacetime.)
- New astronomical objects (there are beliefs that astronomical objects like white holes, quark stars, preon stars, cosmic strings, naked singularities, and so on may exist.)

THE GRAND SCIENCE OF PHYSICS

- Life on other worlds (there are beliefs that life exists on other worlds.)
- Alien intelligence (the belief that intelligent alien beings exist.)
- Life on Mars. Its thought this planet may have microbes.
- Life in the Solar System. Its thought Mars, Europa, Titan, and so on may have life.
- The reception of radio signals originating from unknown alien civilizations.
- A unified force (this is the belief in a supreme force of Nature.)
- Antigravity (this is the belief in gravity control or an antigravity force of some kind.)
- Controlled fusion (this is the dream of harnessing fusion in a reactor.)
- The development of a working tokamak reactor (fusion reactor).
- The harnessing of mythical energies like orgone, odyle, Earth energy, etc.
- The discovery that a constant can change, a law can be violated, that a conservation law can be violated, and other epic dramas.
- The development of a quantum computer.
- Gold in the sea. Its known that the world's oceans are filled with tons of dissolved gold. Finding a way to acquire this dissolved gold from seawater has been a challenge.
- Discovery of the "brane", a "membrane" of space.
- Solving the mystery of how the universe will end.
- Solving the world energy crisis problem.
- Room temperature superconductor (there is the dream of making a chemical that superconducts at room temperature.)
- The creation of a room temperature superfluid.
- Supersymmetry (this is the theory research fad of finding new particles in Nature.)
- M theory is an offshoot of superstring theory.
- Supergravity (this is a physics research fad to find a grand unified theory.)

- Proving any of the speculated theory fads in advanced physics are real.
- Dark energy, dark matter (this is the belief that the universe contains mysterious forms of matter and energy that so far are undetected.)
- Shadow matter (this is the belief that a "counterpart" to matter exists that has not been discovered as of yet.)
- Inflation (this is Alan Guth's idea that the universe "super-expanded" as it was born.)
- Paranormal (this is the age-old belief in magical abilities like levitation and telepathy.)
- Above TeV. TeV refers to energies of "trillions of electron volts". There is a belief that perhaps new physics phenomena "above" these energies may be discovered.
- Quark star (this is the belief a kind of star made of quarks exists.)
- Free quark (this is the belief a quark that exists outside the nucleus will be found.)
- Singularity (this is the belief that at the center of a black hole exists a point-size object of infinite mass and density.)
- Relic gravity waves (this is the belief in gravity waves produced by the Big Bang.)
- Condensates (this is the belief in other kinds of condensates like the BEC.)
- Unification of forces (this is the belief that all fundamental forces derive from a single, master force of Nature, so far undiscovered.)
- Cold fusion (this is the idea that "table top" controlled fusion is possible.)
- The creation of a physics genius computer, a creative computer, and so on.
- Nuclear waste problem solution (this refers to proposals of how to get rid of nuclear waste.)

- Rips in space (this refers to spacetime anomalies that rip space in some way.)
- Supersolid (this refers to a novel state of matter occurring near Absolute Zero.)
- New planets, Nemesis (this refers to beliefs that perhaps new planets orbit the Sun or an unknown companion star called Nemesis exists about the Sun.)
- Graviton (this refers to a speculated particle of gravity.)
- New technologies (this refers to the discovery of new technologies to come.)
- Twistor theory (this is a physics research fad into a new kind of theory.)
- Quantum reality (this refers to discussions on the meaning of quantum mechanics.)
- New lasers (this refers to the development of new kinds of laser.)
- Monumental acts (this refers to possible, but unforeseen monumental acts of physics to happen in the future.)
- Other universes (this refers to the discussion on the existence of new kinds of universe.)
- Spaser (this refers to a novel kind of laser that has appeared as of 2009.)
- Nanotechnology (this refers to technology that is of the size of atoms, cells, and molecules. It is a subject improving vastly with a lot of potential for more discovery.)
- The confirmation of some new and radical theory of physics.
- Possible developments out of Albert Einstein's work.
- Its thought the Big Bang created all kinds of exotic physical object like mini-black holes, magnetic monopoles, relic black holes, and more. It is possible some may be found.
- Physics creativity research is the blending of physics with creativity research.
- The use of quantum effects or other physical effects in original ways.

THE NOBEL PRIZE FOR PHYSICS

In all it is not known what will win the Nobel Prize in the coming years. Whatever it is it should be something monumental for physics in a discovery, insight, law, research path, or else. The Nobel Prize is an award with a stature that it commends great acts of physics and for this the physics community expects Nobel committees to award prizes for just that.

Legacy of Nobel

Alfred Nobel by creating his prizes gave the science world an award that glorifies scientists who did epic things. His prizes are today the world's highest honors of science and the people who won them have influenced the world immensely. Today, Nobel's influence affects the world over by his prizes. Consequences of Nobel's influence are:

- Element Nobelium –Nobel museum –Ig Nobel prize –Nobel conferences –Nobel streets –national pride over having Nobel laureates –the prestige of Nobel laureates –"Nobel rushes" to do science work that could win a Nobel prize –dreams of scientists to win a Nobel prize –Nobel lectures –Nobel laureates serving on commissions and panels -Nobel statues -Nobel Institute

In all the Nobel Prize is physics' most eminent and highest science award. It was created to commend persons who did something great, outstanding, monumental, or epic for physics. It serves Nobel's intentions of conferring fame and prominence on physicists who serve physics in a great way.

In the next section, we will now conclude this book with finishing discussions.

"Alfred Nobel did a great service to physics. He established a prize that would glorify the physicist. It would make the physicist seem like a rock or movie star. It gives the the physicist a glory he would not have any other way. His job is to serve physics and for this anonymity is a physicist's fate."

-physics lecture

"Physics is a cause on par with humanitarianism, music, and crusading for peace. It is done for noble ends, to understand the universe, and to abolish superstition as belief of Man."
                                                            -physics lecture

Conclusion

Physics is a vast and complex science. It was born ages ago amidst superstition and natural philosophy. It would go through several ages mired in superstition, but in time it would emerge as its own science. It is an attempt to understand the universe by means of proof, experiment, and reason. For this it has achieved very well. It continues being the epic science it is. In future ages, it will advance to unbelievable complexity more than what this age is. It is a fantastic creation of Man and achievement overall. It is the science of physics, a gift to understand the universe.

Please read widely on physics for your own experience and journey through this epic science.

This completes our discussion on the science of physics and its related issues.

Thank you for reading.

The End.

"Physics is a tough science. Most people will not learn physics and do not understand it. It is the task of the teacher to make it understandable."

-Voigt

About the Author

Tim Vok would major in physics at a college in the US state of Wisconsin. He would go on to a career working in domains like:
- science writing -tour guide -seasonal employment in places like the Grand Canyon, Mt. Rushmore, Bryce Canyon, and Yellowstone
- English teaching -cooking -investing -metal detecting and prospecting -camp counselor -manufacturing -picking

Vok has certifications in the following areas as well:
- English teacher certification -Property management course -Tax preparation

Vok would study physics for over 20 years. He would gain exposure to physics while in high school and watching science television programs. He would take courses in electromagnetism, mechanics, optics, quantum mechanics, and other domains. He has worked as an astronomy assistant, joined several physics clubs, won recognition in Science Olympiad both as a medalist and as a volunteer, gained a college degree in physics and mathematics, and has worked under professional physicists in positions like tutor, lab assistant, erstwhile student, and physics aide. For some reason, he would not go to graduate school and obtain the doctoral degree in physics, thus he would not become a physics professor. His more epic experiences in physics are the dramas:

Writing a book on the search for the mythical theory, the grand unified theory.

Exploring the subject of UFO research and writing a popular science account of this amazing and controversial subject.

Designing video game concepts where physics is taught in an entertaining way in association with adventure video games.

Writing books on all manner of science issues.

He would fantasize and invent several kinds of bomb in science fiction like the 'Big Bang bomb', world bomb, and galaxy bomb.

He would add prefixes to physics ideas and derive science fiction ideas like the following:

- antispace -antienergy -subtime -antiforce -hyperworld -post universe -superverse (super-universe) -hypermatter –hypertime -anti-spacetime -hyper-spacetime -hyper-energy -supermatter -protospace -prototime -sub-energy -untime -unspace -unmatter

Helping found the newly emerging physics discipline known as 'physics creativity research'.

Proposing names for newly discovered dwarf planets and new moons of the dwarf planet Pluto.

In 2012, new moons of Pluto would be discovered and he would propose naming one of them as Cerberus, the three-headed demon hound of Greek myth.

In 2005, he would propose the name of 'Concord' for the new object, 2003 UB 313, later renamed as dwarf planet, Eris.

Vok would visit sites famous in physics lore like Lowell Observatory (site of the discovery of Pluto in 1930).

Vok would propose making a computer capable of creativity and doing physics as well.

Vok would invent particle physics' imaginary particles like the 'smalleston' (smallest particle of all) and 'infinitesimon' (infinitely smallest particle).

He would propose to the NASA space mission known as WISE an idea of searching for the mythical companion star of the Sun, Nemesis. Its believed this NASA mission would in its research survey objects

ABOUT THE AUTHOR

unknown near to the Sun. There is a long standing mystery in astronomy that the Sun has a companion star and this is dubbed Nemesis. It is presently undiscovered, but its thought the WISE mission would have discovered this by now, but has not recognized it as of yet. If it is found, then this would present a titanic moment for astronomy and the solar system as the Sun has a companion star and is a binary system. He would also propose that it may have discovered an unknown planet orbiting in the outer solar system. There is strong evidence the Sun has another planet in its system, but it is presently unknown as of now. If these objects exist, they await for scientists of other times to be lucky enough to find them.

Vok would work at Mt. Rushmore on occassion. He would propose making a grand monument to physics by sculpting a small mountain for physics.

In all Tim Vok feels it was a pleasure and a gift of life to study physics and make it a passion. It is a fascinating science enriching any student who majors in it. It is a a fantastic science where the student can learn about the nature of the universe. Vok encourages all students to study physics at some time in life. Vok is a fan of the TV show, Carl Sagan's COSMOS. Vok encourages people to read Carl Sagan's books and to study his monumental life in science.

Please sample the many kinds of e-book written under names: Tim Vok and Tim Votk.

He presently lives in Arizona in 2012.

"Physics is a journey. It is not like climbing a mountain, though it can be. It is fascinating and mysterious, addictive and compelling. The traveler wants to learn physics. Each moment is its own adventure and the journey of it is a great reward."

-Voigt

www.ingramcontent.com/pod-product-compliance
Lightning Source LLC
Chambersburg PA
CBHW020724180526
45163CB00001B/94